Praise for *Future Memory*

"This knowing and this perceiving, which is outside of space-time, is non-local and all pervading. It opens the door to an interconnection of all life. *Future Memory* is a most intriguing book. Thousands of people have now had important glimpses of higher states of consciousness through near-death experiences (NDEs), and Atwater is one of the investigators of these. The ideas she has put together from her own and others' experiences here will help a lot of people make more sense of profound human experiences."

—Charles T. Tart, Ph.D., author of *Altered States of Consciousness* and *Transpersonal Psychologies*

"This is a book appropriate for the type of consciousness exploration we are involved in today. As we continue to be receptive to and involved in the unfoldment of quantum physics, we see more clearly that consciousness is the foundation of what we know and perceive."

—International Association for Near-Death Studies' *Vital Signs* magazine

"*Future Memory* is a very interesting book. Brilliantly written, east to read (even for me), with tons of scholarly information, scientific wisdom, and spiritual clues—about timeless time, dimensionless space, and boundless mind. Every chapter is worth reading."

—Tulku Thondup Rinpoche, author of *Hidden Teachings of Tibet*

"Atwater has taken a major step past her earlier work, *Beyond the Light,* rather laying challenge upon challenge to our mind and spirit. May she be heard far and wide."

—Joseph Chilton Pearce, author of *The Crack in the Cosmic Egg* and *Magical Child*

"As well researched as it is provocative, this book questions the very nature of time, memory, and the scope of human consciousness. *Future Memory* is unlike any other book you will ever read."

—Alfred Dolezal

Books by the Author

As You Die (narrative on CD or DVD)

Beyond the Indigo Children:
The New Children and the Coming of The Fifth World

Beyond the Light:
What Isn't Being Said about the Near-Death Experience

The Big Book of Near-Death Experiences

Children of the Fifth World:
A Guide to the Coming Changes in Human Consciousness

Children of the New Millennium

Coming Back to Life

The Complete Idiot's Guide to Near-Death Experiences

Goddess Runes

I Died Three Times in 1977 – The Complete Story

The Magical Language of Runes

Near-Death Experiences: The Rest of the Story

The New Children and Near-Death Experiences

The Runes of the Goddess (book and kit)

We Live Forever: The Real Truth about Death

E-Books: Life Sounds

The Frost Diamond

The Challenge of September 11

I Died Three Times in 1977 (original four articles)

FUTURE MEMORY

P.M.H. ATWATER

HAMPTON ROADS

Cover design by Jim Warner
Cover art by © Ocean/Corbis
Interior designed by Frame25 Productions

Hampton Roads Publishing Company, Inc.
Charlottesville, VA 22906
Distributed by Red Wheel/Weiser, LLC
www.redwheelweiser.com

Sign up for our newsletter and special offers by going to
www.redwheelweiser.com/newsletter/.

ISBN: 978-1-57174-688-7

Library of Congress Cataloging-in-Publication Data available upon request.

Printed on acid-free paper in the United States of America

TS

10 9 8 7 6 5 4 3 2 1

Dedication

This book is lovingly dedicated to the three special people whom I was privileged to bring forth in birth:

Kelly John Huffman—a child of the oceans with a heart of music who seeks to learn harmony from life.

Natalie Gae Huffman—a child of fire with the heart of a gypsy who seeks to learn courage from life.

Pauline Ann Huffman—a child of night with the heart of a rosebud who seeks to learn wisdom from life.

Contents

Acknowledgments

The book you are about to read constitutes the most extensive research project I have ever conducted. Because of this, many people have been directly or indirectly involved in its making. Those who contributed the most are listed below, yet there are thousands more. To each and every one of them, I offer my sincerest and deepest thanks!

Terry Young Atwater
David Morgan
Vic Bertrand
Dennis Swartz
Stephany Evans
Steve Sommerfeld
Liz St. Clair
Peter R. Rothschild
David McKnight
Mirtala
Nic and Estill Tideman
Rita and Noel McInnis
Walter and Mary Jo Uphoff
Patricia Helen Home
Walter Starcke
Machaelle and Clarence Wright
Sun Bear and Wabun Wind
Diane Pike and Arleen Lorrance
Elisabeth Kubler-Ross
Lao Russell

Don and Neddy Repp

Arthur E. Yensen

William G. and Jeanie Reimer

Kenneth Ring

M. Elizabeth Macinata

Terry Macinata

Thomas More Huber (Thomas Shawnodese Wind)

Charles and Lee Wise

Jack Huffman

Johnny Lister

Rev. Paul Neary

Roger Pile

Annie Sargent

Art Samson

Oralee Stiles and Marzenda Stiles McComb

The people at Idaho First National Bank (now West One) and especially Sandi Bonnett Pack, Karen Woods, Kathy Pidjeon, and Janet Atkinson.

Introduction

For anyone glued to the planet's surface by gravity, the expression "high" is an adjective whose meaning is obvious. Yet, when the forces of gravity are counteracted or become too weak to act on a perceivable mass, the meaning of high topples hopelessly out of context. In "free fall"—a state only recently experienced by space-faring humans—gravity apparently disappears and the notion of up and down turns into a comfortably "user-definable" variable.

You see, if we boil all facts down to brass tacks, space does not even exist. For space, as such, is but an auxiliary concept that enables us to relate one object to another. Without space, we could never master chaos. Space is essential for our growth. It seems obvious that if there were absolutely no objects—things—space would be totally meaningless.

The same holds true for time, which is another auxiliary concept of paramount importance. Time's purpose is to relate events to one another. Should absolutely no event ever occur, time would be completely void of significance; thus it wouldn't even exist. The fewer the events, the looser the division of time. In biblical times, when during long millennia events unfolded at a rather slow pace, the smallest time segments were represented by thirds of the day: morning, afternoon, and night.

The need for the hour did not arise until the so-called Middle Ages. Minutes were not required until much later, when the pulse of events began to hasten and shorter time segments became a necessity. Seconds appeared at the end of the 1800s. But a few decades ago, when human interference in subatomic events began to require speeds that only supercomputers can handle, the urgent necessity for high-precision time measurements developed.

Shortly, it became evident that the concepts of space and time were hopelessly unmanageable without relating both to one another. So, early in the 1900s, a new breed of hard-core philosophers—the scientists—coalesced these concepts into a bafflingly manifold notion called space/time.

The revolution triggered by formulations derived from the space/time concept revealed a fantastic, hitherto unimaginable branch of physics—quantum mechanics—that affected even religious thought, because, for the first time, sophisticated equations could not be validated without the recognition of God. The new sciences appeared to offer an equation for everything. But do they, really?

P.M.H. Atwater offers surprising—and, yes, even awesome—answers to this question in her, both bold and delightful masterpiece entitled *Future Memory*. It is, in effect, a highly practical treatise that takes you by the hand and walks you through a fascinating array of phenomena that has perplexed our human society since it began to develop. As a matter of fact, she guides you unerringly through undeniable portents that become more and more unsettling as civilization escalates ever newer summits.

Future Memory is dedicated to the thinker—which I believe you are, Gentle Reader, for if you weren't, you would not be reading these lines—and it will make you feel elated when you discover that wherever you turn, you are a fundamental participant in the processes of Creation.

Regardless of anyone's point of view, the universe is eminently and inalterably hierarchic. Everything in the Cosmos is subject to hierarchic orders for the simple reason that nothing could ever interact with anything if it were not for the orderly and logical mechanisms governed by hierarchy's majestic laws. Human society represents one of the most eloquent examples of this truth.

The very stability—and, consequently, the perpetuity—of the universe depends on the coherent interaction of its components. This means there is no universal equality. The idea of an equalitarian society is but a widely accepted myth. Equality, if you appreciate the pun, exists only among equals. Unfortunately, the majority of the people in existing societies are not aware of their true identities, which explains, at least in part, the prodigal multiplicity of religions and cults. The scenario is completed by the infectious trend of hero worship, fads, and an impressive array of all sorts of malarkey instilled into our consumer society by the media. The mediocre is being acquainted with his mediocrity, the slipshod is shown a mirror of weakness revealed.

So, where does all this leave us? The answer is simple enough. As we have seen, hierarchies are part of nature. We cannot fight anything that is natural. Nonetheless, the fact that you have evolved to a height at which you become consciously aware of this implies unequivocally that you have successfully escalated the higher ranks of the hierarchy.

And this is what *Future Memory* is all about. P.M.H. offers you a highly effective and exciting crash-course in discovering and exploring the truth of your real identity. She will teach you who you are and how to access your Maker, without endorsing intermediaries or resorting to outrageous imbecilities. If you put her teachings into practice, you will discover yourself and your place at the side of the Creator.

After you have finished reading *Future Memory,* you will have discovered the high road to ultimate happiness. And you will rejoice to know that you really are an equal among true equals.

Peter R. Rothschild, M.D., Ph.D., D.D., F.R.S.H.
Nominated for the Nobel Prize in Physics, 1986

The Purpose of This Book

Physicist Wolfgang Pauli once decreed that a new science is needed to explore the objective side of human consciousness and the subjective side of matter. Not mysticism, but a science willing to incorporate objective and subjective avenues to discovery while recognizing the legitimacy of personal experience. He realized that intimate encounters can give shape and quality to scientific models, enabling multiple layers of information to surface.

Pauli's decree gave me the courage I needed to write this book, an attempt to describe the indescribable—that moment of brain shift after a spiritual transformation when realities switch and new ways of perceiving existence replace former beliefs.

During the four decades I have researched near-death experiences and other transformative states, I discovered that one of the aftereffects of such a shift in consciousness is what I have come to call the "future memory phenomenon," a peculiar condition whereby people are able to live life in advance of its physical manifestation and remember having done so when something triggers that memory. I used this discovery to launch a much deeper, broad-based exploration of consciousness and its link with matter and energy and light. My goal in doing so was to construct a framework, hopefully a meaningful context, with which we might better understand what happens during a transformative brain shift . . . and . . . what may be at the very core of existence itself.

I tell myself this is a research book, the most extensive project I have ever undertaken, a veritable adventure story wrapped around both objective and subjective ways of perceiving reality. Yet, in truth, this book is much, much more.

The real purpose of this book is to provide a way for anyone to experience a shift in consciousness *simply by reading the pages that follow*. This can and will happen if you stay on the path. Don't skip-read or read the end first. Start at the beginning and go straight through. If you don't, the book will not make sense, nor will the material hold together. That's because *Future Memory* is not a book. It is a labyrinth, a real one. Like with site labyrinths, directions will change again and again, coiling around entrance to a center point, then uncoiling as you circle back through where you've already been—only this time seeing yourself and the world around you in a new way.

It doesn't matter if you understand everything. Just the experience of reading through a labyrinth in the same manner as one would walk through one is very healing. Any type of labyrinth does this, enables consciousness to rise to its next highest level—as if freed from constraints. Labyrinths can be revisited, rewalked, in this case reread, for another lift, another opportunity to shift again.

That makes the book you are now reading a psychotronic device, a literal brain-changer. Don't credit me for this unique design. During my third near-death experience (I had three in three months in 1977), a voice spoke to me, a voice so big and so powerful it filled the entire universe and all of Creation. I called it The Voice Like None Other. My sense was it was of God. The Voice told me to do the research I have done and to write three specific books, naming this one as book two. Not until later was I shown how to accomplish this. That "show and tell" happened after I had made a fool of myself writing seven different versions of this book, none of them interesting a single publisher or passing muster with reviewers.

I was ready to burn the thing when, after a shower, suddenly all motion froze, space expanded, and the air filled with super-bright sparkles. For the second and only other time, The Voice spoke, showing me how to redesign each sentence, each paragraph, each page mathematically in the format of a labyrinth . . . a book labyrinth. It took me two weeks. Days later the book was grabbed by Birch Lane Press and published in 1996. Hampton Roads is now reissuing it. Thank you to my editor, Greg Brandenburgh, for having the vision to do this.

It doesn't matter what I tell myself about *Future Memory*. The fact is this book is *alive* in how it affects the brain/mind assembly and changes people, in how it helps to facilitate an actual brain shift. You don't just read the book. You uplift and expand your own consciousness as you explore the coiled passageways it provides throughout the wonder of your own brain.

Enjoy where it takes you.

The Five Blind Men and the Elephant

One day five blind men decided to find out what an elephant "looked" like. Led to one, each man grabbed hold of whatever section of the animal he could, certain that what he had grabbed was the whole of the elephant itself.

The one holding onto the trunk thought for certain that the elephant must indeed be the shape of a wiggly snake and said so, but the one who had found an ear countered, insisting that the elephant's size was that of a palm leaf. With a firm grip on one of the elephant's legs, the third announced that the animal was actually like a tree trunk. "No," stated another, while patting the elephant's side, "this beast is truly the size of a wall."

Then the fifth, being the loudest and most impatient of the group, clasped the animal's tail and yelled, "Oh, my brothers, you are not only blind but crazy, for the elephant is the shape of a rope."

—Ancient Hindu parable

Part I

Future Memory

1

The Labyrinth Begins Here

You can't cross the sea merely by standing and staring at the water.
—Sir Rabindranath Tagore

Journey with me through the universe of the mind, into deeper realms of internal and external environments, where states of consciousness play out like overleafs upon the backdrop of personality and place—who we think we are and where we think we live.

Few ever question these deeper realms, much less bother to investigate them. We do, and that's what this book is about.

The ancient Hindu parable of "The Five Blind Men and The Elephant" best defines the territory we are about to tackle: the span that exists between perception and truth, between what seems real and what is real, between life's many puzzles and how they interconnect and interweave.

We tackle this territory for one reason, to search for what neither perception nor truth can supply—perspective—the perspective to understand why existence exists and why we are who we are. And we do this, attempt to describe what is thought to be indescribable, in a spirit of high adventure.

To begin our journey, we will explore shifts in the awareness of reality, among them future memory—a peculiar phenomenon that challenges our understanding of sequence. Like the old riddle "Which comes first, the chicken or the egg?" we will question the necessity of one event to always follow on the heels of another. In part two, we will grapple with the inner workings of consciousness and especially how that relates to creation itself, the universe, planets, souls, and the concept of deity. This will enable us to see how time and space can be but mere illusions in a grander scheme of life after life. Finally, in the last section, we will step beyond notions of real versus unreal to confront the truth that undergirds existence itself.

A common thread interweaves our journey—what the ability to remember the future reveals about brain development.

A common admonition fuels our passage—know thyself, for knowledge without wisdom can distort and deceive.

A common desire ever directs us toward our goal—the reawakening of wonderment.

As we embark upon this, the journey of a lifetime, consider first the following observations summarized from various studies conducted over the years on early childhood development.

Children prelive the future on a regular basis. By the age of four, the average youngster spends more time in the future than in the present. The temporal lobes of the brain develop during this period, enabling the child to project ahead and rehearse in advance whatever might someday be expected of him or her. Children play with futuristic possibilities and potential outcomes as a way of "getting ready." A child's preoccupation with the future is healthy. It is a natural component of growth, a desirable state of affairs ensuring that both brain structure and brain capacity meet the requirements of *an emerging consciousness.*

What if adults do the same thing as children once their brains shift after a transformational event such as a near-death episode, a shamanic vision quest, a kundalini breakthrough into spiritual enlightenment, a religious conversion, or because of head trauma or being hit by lightning?

What if the adult ability to prelive the future is actually *a reliable signal that temporal lobes are expanding*—so an increase in brain structure and brain capacity can be accommodated, preparatory to accessing *enlargements of consciousness*?

It is known that people who experience major spiritual transformations become more childlike afterward, in the sense that they often glow with a newfound innocence plus a desire to relearn and redefine life. Most of them possess levels of curiosity and intelligence greater than before, and are seldom affected by limitations from former attitudes and beliefs. Able to easily slip in and out of stages of behavior development once thought the exclusive domain of youngsters, such experiencers appear to "grow up" all over again.

During the twenty-plus years I spoke with or interviewed thousands of near-death survivors and their families and friends, I noted that the ability to "remember" the future was quite typical of the aftereffects. Most experiencers displayed the trait. Yet I also observed this same characteristic with people who had never undergone a transformational event of any kind. This so surprised me that I sought out these "other" people.

What I found challenges how brain development is viewed and how "real versus unreal" is determined. Indicated as well is the distinct possibility that each and every one of us may be able to transcend our daily fare and quite literally live the future before it occurs. (Don't confuse what I am saying here with what some near-death researchers term "flash forwards." What I have observed and experienced myself is far more complex and dynamic than that, physically real to the individual involved and lived in minute and verifiable detail.)

It may strike you as odd the way I've decided to arrange this book and divulge my findings, yet what follows is the only viable framework I could construct that can encompass the enormity of the information we need to cover while at the same time providing us both with a fun trip.

Yes, I turned this book into a labyrinth, one you can traverse via the written word. Throughout its pages, I intend to tease you with a labyrinth of subjective and objective stories and facts so that the awesome wonder of what lies beyond worlds internal and external to us can emerge.

What do I mean by a labyrinth?

According to the dictionary a labyrinth is a devious arrangement of passages and pathways that form the pattern of a maze. No ordinary maze, admittedly, but one with a single way in and out, encircling as it enfolds back upon itself, again and again, in steady progression toward a central core.

According to tradition, however, walking, running, or dancing one's way through a labyrinth invokes a sense of healing and balance in the participant—order out of chaos, if you will. That's because a labyrinth is designed to stimulate the expansion of a higher form of consciousness. A typical maze is meant to confuse; but a labyrinth, with spirals that mimic the convolutions of the brain, leads one into the depths of soul, arousing a gut response to the mysteries of creation, of birth and life and death and rebirth. You're turned "inside out," and memory, the memory of who you really are, is awakened.

To quote Reverend Dr. Lauren Artress, author of the charming book *Walking a Sacred Path: Rediscovering the Labyrinth as a Spiritual Tool*: "Walking the labyrinth clears the mind and gives insight into the spiritual journey. It urges action. It calms people in the throes of life transitions. It helps them see their lives in the context of a path, a pilgrimage. They realize that they are not human beings on a spiritual path but spiritual beings on a human path."[1]

This book is a labyrinth in the truest sense. Each chapter and section has been crafted to both stretch your mind and challenge your beliefs. You read your way through it in the same manner as you might walk through a labyrinth, and with the same result—a shift in consciousness.

Take a deep breath.

Ready yourself.

Your exploratory journey through the universe of time and space and memory and consciousness—through the enfolding turns of the labyrinth—has begun.

A left-handed labyrinth	*Christian eleven-turn labyrinth*	*Minoan Cretan seven-turn labyrinth*

2

Reality Shifts

We human beings invent reality as much as we discover it.
—Lawrence LeShan

Perception determines "truth." We invent our own reality this way, by not questioning perception, ours or anyone else's, and by accepting what appears to be real as "real."

History is filled with stories of people who, in "slipping between the cracks" of their own consciousness (thus altering how they perceived the world around them) uncovered different ways to experience reality. What they accomplished in doing this made an impact on society. You and I, all of us, have profited again and again and still are profiting because this happened, because some people were dazzled by wonder instead of silenced by dogma.

Here are some examples of what I mean. Tripping through these tidbits will help establish common reference points, which are necessary, I believe, for the exploratory journey we have embarked upon.

Let's start with Xerox.

Chester F. Carlson, inventor of the Xerox duplication process and founder of the Xerox Corporation, was a devotee of a certain trance medium who channeled spirit beings from The Other Side. While attending a series of sessions with the woman, he eventually "received" the photocopy process from the spirit beings she contacted. After experimenting with the technique and making a few adjustments, Xerox was "born" along with a multibillion-dollar company. I don't know about you, but after having worked as an executive secretary during the fifties and sixties and making the copies I needed the only way you could then—by stuffing packs of carbon paper in a typewriter and banging the keys hard enough to make the images pass through—I am immensely grateful that Carlson frequented seances. He revolutionized the business world because he did (and certainly made my life easier).

Switching to peanut butter... the *Woodrew Update,* a well-researched and provocative newsletter,[2] reminds us that:

> George Washington Carver took the peanut, until then used as hog food, and the exotic and neglected sweet potato, and turned them into hundreds of products. His list included cosmetics, grease, printer's ink, coffee, and, of course, peanut butter. Carver said he got his answers by walking in the woods at four in the morning. "Nature is the greatest teacher and I learn from her best when others are asleep," he said. "In the still hours before sunrise, God tells me of the plans I am to fulfill."
>
> Thomas Edison unsuccessfully tried to hire Carver, confiding to his associates, "Carver is worth a fortune." Henry Ford also tried (and failed) to hire him, calling Carver "the greatest living scientist." When Carver's plans were fulfilled and his dreams translated into realities, he refused to take out any patents, believing that all inventions and discoveries belonged to mankind, not to one man. The result is that much of what he learned is lost in the annals of history.

How did George Washington Carver communicate with God during the wee hours of morning? He said it himself: through the assistance of angels and fairies. And he isn't the only one to make such a claim. Peter and Eileen Caddy and their colleague Dorothy Maclean give the same credits in describing the work they accomplished.

This troupe, along with Caddy's three sons, took up residence near an inlet to the North Sea at Findhorn, Scotland, for the purpose of setting up a cocreative two-way link between themselves and nature intelligencies—that is to say, angels (what they later called "devas") and fairies ("nature spirits"). They became willing workers with nature's own in an attempt to cocreate a garden the likes of which would defy every known rule of convention and climate. That was 1962. Today, the now famous Findhorn Gardens regularly draws people from across the globe to tour the premises and take classes at Cluny Hill College, classes on how to communicate with angelic forces and helper spirits, while at the same time enhancing one's own sense of spirituality.[3]

Perelandra Gardens, the next generation in angel-fairy communication, goes beyond that of a fixed-based operation (like at Findhorn) to a method anyone can use no matter where they live to improve their life and expand their consciousness. Started by Machaelle Small Wright and her partner, Clarence Wright, this forty-five-acre hideaway about fifty miles south of Washington, D.C., near the tiny hamlet of Jefferson-ton, has grown from the initial experience in 1973 of Machaelle hearing disembodied "voices" coming from the woods on their property to a well-established Center for Nature Research, an open-air laboratory dedicated to the discovery of nature's laws and the principles and dynamics behind the cocreative relationship between humans and the varied intelligencies of nature.[4]

The people I have mentioned came to perceive reality from another vantage point; then they used what they gained from that experience to benefit others, as this addendum to the Perelandra story demonstrates:

> According to a Canadian agronomist, soil samples taken from Perelandra in 1989 tested out with the highest vitality rating of any soil ever tested—until several other people achieved the same rating that same year using the Perelandra Method on soil in their own gardens. For this reason, and because of extensive documentation on methodology and results, the internationally esteemed magazine *Organic Gardening* featured an article about Perelandra in their

November 1990 issue. The article quoted Albert Schatz, Ph.D., a retired professor at Temple University in Philadelphia, and the one who introduced the concept of chelation (a major factor in soil formation), as saying: "I have spent a lot of time trying to find scientific holes in what Wright is doing and I have not been able to do that, which amazed me. She has integrated health and agriculture in a very unique and effective way, and her research is scientifically sound and valid. I can find no evidence to draw any other conclusions. Her research is the basis for the future of agriculture."

Different ways of experiencing reality happen when individuals expand their consciousness. Whether accidently or on purpose, that shift in perception also alters the validity and the importance of time and space.

Documented cases of native runners, especially those in North and South America, illustrate this. The Spanish, for instance, once recorded native runners who could cover in excess of 150 miles per day, making the trip from Lima, Peru, to Cuzco in three days, where it took Spanish riders on horseback twelve days to do the same thing. Running was and still is considered a sacred task by native peoples. Adherents observe strict disciplines in exchange for what they believe are the holy gifts of speed and invisibility.

From Peter Nabokov's book *Indian Running,* an anthropologist by the name of George Laird described what happened to a particular runner who lived in the southwestern part of the United States: "One morning he left his friends at Cotton Wood Island in Nevada and said he was going to the mouth of the Gila River in southern Arizona. He didn't want anyone else along, but, when he was out of sight, the others began tracking him. Beyond the nearby dunes his stride changed. The tracks looked as if he had just been staggering along, taking giant steps, his feet touching the ground at long irregular intervals, leaving prints that became further and further apart and lighter and lighter in the sand. When they got to Fort Yuma, they learned that he had arrived at sunrise of the same day he had left them."[5] (Thus arriving before he departed.)

We know native peoples are capable of such feats because so many cases have been studied. But let's not forget the Australian aborigines. Theirs is the oldest continuously existing culture on earth (around for at least fifty-thousand years), and they maintain an understanding of time and space, of reality, that deserves our attention.

What they call "dreaming" has little to do with sleep or dreams that occur during sleep. Dreaming for them is actually more akin to a type of "flow," where one becomes whatever is focused upon and suddenly knows whatever needs to be known at the moment. Aborigines sometimes use drugs to achieve this state but, more often than not, drumming, chanting, rhythmic movements, certain sounds, and rituals suffice. In this state of consciousness, participants seem to "merge with" or "enter into" soil, rocks, animals, sky, or whatever else they focus on—including the "In-Between" (that which appears to exist between time and space, as if through a "crack" in creation).

Of interest is the fact that these people believe reality consists of two space-time continua, not one—that which can be experienced during wake time and that during dream time, with dream time slightly ahead of its counterpart, yet capable of merging into "all time," or what Pulitzer Prize-winning poet Gary Snyder calls "everywhen." To Australian aborigines, wake time is where learning is acted out and utilized, but dream time is where learning is first acquired. For them, dream time is the place where all possibilities and all memory reside. Stories are told even yet of aborigines who physically appear and disappear as they slip back and forth from one continuum to the other, and alternate universes they believe exist and the everywhen they know awaits them.

Wise ones, be they monks or shamans or healers or mystics, are like this. They know life extends beyond the boundaries of perception. Yet, as the parable of the five blind men and the elephant makes clear, perception itself can be flawed.

Yes, it is a fact that individuals and societies have always organized the cosmos to fit their own preferred beliefs. This is what defines the relationship between heresy (independent thinking) and orthodoxy (mutually accepted bias). But it is also a fact that the bizarre can intrude upon one's life so dramatically that one is forced to shift one's awareness of real versus unreal.

Reality shifts (sometimes called "coincidences") take on many guises. Samples of the "fictional" kind:

- The popular movie *China Syndrome*, starring Jane Fonda, depicted a nuclear facility meltdown *three weeks before the same kind of disaster actually happened* at Three Mile Island near Harrisburg, Pennsylvania.

- The 1961 novel *Strangers in a Strange Land*, written by Robert A. Heinlein, told the story of a global chief executive who made decisions based on his wife's advice, advice she had obtained from regular consultations with a San Francisco astrologer. In 1988 media headlines carried the story that Nancy Reagan frequently consulted with a San Francisco astrologer, and that the advice she passed along to her husband Ronald Reagan, then president of the United States, *was based on those consultations.*

- The novel *Futility*, an 1898 creation of Morgan Robertson, detailed the sinking of an unsinkable ship by the name of Titan, the largest vessel afloat. This imaginery ship collided with an iceberg during April, resulting in a high loss of life because the *Titan* carried too few lifeboats. Fourteen years later the real ship *Titanic* recreated what happened in the novel with uncanny similarities: The two ships had almost identical names, both were designated unsinkable, both were touted as the largest ships at sea, both collided with icebergs in April, both suffered high human losses because of too-few lifeboats ... *plus*, both had strikingly similar floor plans and technical descriptions.

- The famous broadcaster Paul Harvey, on his radio show *The Rest of the Story*, aired a grim tale of three shipwrecked sailors and one cabin boy, adrift and facing starvation, who drew lots to see who would forfeit his life so the others could survive. The contest was rigged to make certain the cabin boy, Richard Parker, would lose. Evidence used at the subsequent court trial after the men were picked up, incontrovertible evidence that convicted all three of murder and cannibalism, *was a story written by Edgar Allen Poe* entitled "The Narrative of Arthur Gordon Pym of Nantucket." In it, Poe told of three shipwrecked sailors who rigged a drawing of lots, then killed and ate their cabin boy companion, Richard Parker. Poe's story, which so accurately described the drama, every detail as it actually happened—including the victim's correct name—was written and published *forty-six years before the event ever happened*, even *before* the participants were born.

The astonishing ability of fiction to accurately foreshadow what physically occurs happens more often than you might think. It's almost as if on some level, knowingly or unknowingly, consistently or occasionally, individuals can tap into or stumble across other dimensions of reality, as well as the ready availability of a predestined or potential future.

But what if the reality shift that occurs cannot be correlated with any sort of imaginings? Samples follow of the nonfiction kind:

- Brad Steiger, in his book *The Reality Game and How to Win It*,[6] tells about Charles W. Ingersoll of Cloquet, Minnesota, who was filmed leaning over the rim of the Grand Canyon, taking pictures with his 35mm camera. The film he appeared in was a travelogue made and copyrighted by Castle Films in 1948. Ingersoll, who had *hoped* to visit the Grand Canyon in 1948 but couldn't, finally made his first trip there in 1955, taking with him a newly purchased camera manufactured the same year of his trip. A week after his return, he chanced upon the old travelogue in a store and bought it, later discovering to his utter amazement that *the film clearly showed him there in 1948— holding a camera that did not exist until 1955.* An investigation verified the incident and the dates, but no explanation was ever offered as to how Ingersoll could have appeared in a film showing him at a site *seven years before* he ever got there.

- On October 21, 1987, Claude and Ellen Thorlin were sitting at breakfast. Ellen *heard* a disembodied voice ask her to tune in Channel 4 on their television set. When she did, there appeared the face of their dear friend and colleague, Friedrich Jergenson, at the moment of his funeral service 420 miles away, a funeral both had wanted to attend but couldn't. Jergenson, a well-known Swedish documentary film producer, is considered the father of the Electronic Voice Phenomenon (obtaining voices on tape or film from disembodied beings). He had spent nearly thirty years of his life recording what he believed to be voices from the dead.[7] Claude Thorlin took a photograph of Jergenson's

TV image and recorded the time at 1:22 p.m.—twenty-two minutes after the funeral service had begun and *appearing on a channel that did not receive broadcast transmissions in the Thorlin's area*. The photograph he took matches what Jergenson looked like near the time of his death.

- When T. L. of Fort Worth, Texas, was twenty-one years old, he borrowed his parent's car for a drive from Darby to Missoula, Montana, to be with friends. Staying later than expected, he found himself driving back to Darby between one and two in the morning at every bit of seventy miles per hour. At a place in the road where it wound around hills paralleling the river channel, the car's headlights suddenly picked up a herd of twenty to thirty horses sauntering across the highway. With no time to hit his brakes and no place to pull off the road, T. L. *hoped* to avoid a collision by driving in between the animals. Two large horses stopped directly in front of his path. The inevitable seemed his fate until, in the flash of an instant, T. L. found himself well beyond the herd, driving as if nothing unusual had happened. To this day he cannot explain how he missed hitting the horses. "It was as if I and my car were 'transported' to the other side of the herd."[8]

Reminiscent of the once popular television series *Outer Limits*, each of these "coincidences" involved people as real as you and I, on what always began as an ordinary day.

Are these events really coincidences? Or happenstance?

Too much evidence exists, collected from too many reports, for either term to apply. Something else is going on here.

Of the reality shifts I have shared with you, note that those incidents that stemmed from the creative process employed an imaginative use of mind, the kind necessary when writing "fictional" literature (illustrating the unlimited scope of altered consciousness). Then notice that the episodes from "real" life hinged on a deep emotional need to fulfill a strong desire (emphasizing the surprising power of emotion).

Both of these types of events underscored moments when *subjective reality overlaid objective reality to determine experience*. And when that happened, *the future easily surfaced . . .* either ahead of time or as a passageway in between time.

This peculiarity occurred automatically, without provocation, and regardless of logic. What we call time—past, present, future—ceased to be sequential for these people and took on the aspect of simultaneity (everywhen).

The preceding cases, all of them—whether involving native or modern-day societies, fictional or nonfictional themes—centered around men and women who encountered alternate versions of time and space once they became open and receptive to possibilities beyond the ordinary. What actually occurred changed their perception of the manifest world. It also changed their awareness of "future."

3

Modes of Futuristic Awareness

Miracles do not happen in contradiction to Nature but only in
contradiction to that which is known to us about Nature.
—Saint Augustine

To experience reality differently is to experience time and space as if both were but a mere illusion. Sequential ordering and causality cease to be meaningful.

You can understand this if you realize that, once consciousness alters to the degree that other realms or dimensions can be accessed, the first thing you lose is a sense of time, the second is a sense of space. The world reorders itself—sometimes significantly—even to the point that the future is capable of unfolding in the present.

The act of awareness is a key we can use to investigate this shift.

Since we want to explore future memory and what it reveals about greater realities and deeper truths, let's take a moment to focus on the traditional modes of awareness people use to obtain futuristic information (either at will or spontaneously). Then, let's look at incidents in the lives of several people that illustrate these awareness modes. I intend this as an exercise to set the stage for identifying the phenomenon of future memory itself and the nature of its characteristics.

Modes of Futuristic Awareness

Prophesying—to predict in advance (*subjective/intuitive*). Style can be psychic or religious. Depends on interpretation of subtle impressions and emotional promptings to provide guidance or advice.

Forecasting—to predict in advance (*objective/logical*). Style is business-like. Depends on mathematical projections made from detailed facts and figures to provide information.

Precognition—to know in advance (*subjective/feeling*). The act of knowing or feeling the future before it happens; occasionally called "sensing." Refers to advance knowledge suddenly known without precursory promptings or impressionistic stimulus of any kind.

Clairvoyance—to see beyond sight (*subjective/seeing*). The ability to clearly see detailed objects, people, or activities not present to the senses, in ways above and beyond natural viewing; an unexplainable extension of eyesight to include images and sights not part of the existent dimension. Refers to being able to see the future happen before it physically occurs, as if an observer to the event.

Clairaudience—to hear beyond sound (*subjective/hearing*). The ability to clearly hear or listen to definite voices and sounds not present to the senses, in ways above and beyond natural hearing; unusual extension of the audio range to include tones and pitches not part of the existent dimension. Can refer to hearing messages about futuristic occurrences or otherworldly activities.

Actually, no matter how far back in history you go, there are records of people who have been privy to the future before it is physically manifested. The information these people gained from this awareness was accurate more times than it was inaccurate. Some of them claimed that such revelations were absolute; others said the future could be altered if people were willing to change certain attitudes and behaviors. Destiny was thus viewed as fixed or flexible—depending. The same is claimed today.

To illustrate these modes of awareness so the characteristics of each can be readily identified, here are several incidents in the lives of two people. One account centers around a life-threatening accident; the other concerns anesthesia given during oral surgery. Utilizing the list of modes you have just read, notice the various ways the following two people became aware of a future reality while still an active participant in present reality.

A Naturally Occurring Experience

Anna Grace Foster was installing a sculpture exhibit in the gallery of Williamsburg Regional Library's Art Center in Williamsburg, Virginia. She had done this countless times without incident, but on this occasion an unpleasant surprise awaited her.

"I had just removed the glass shelves from a heavy movable glass case and turned away to put one down. As I turned my back, the 250-pound case toppled over on me, sending glass shards in all directions and driving me down with its weight.

"Caught like a turtle in its shell, I screamed as loud as I could for attention, since there was no one in the gallery at the time. Two male staff members from elsewhere in the library came running, as did several women from other offices. As the men lifted the case from my body, cautioning me not to move on the bed of glass, someone yelled, 'Dial 911,' the emergency squad number.

"At that moment a clear message went through my head—'I am totally protected by the Holy Spirit, this is a neutral event.' As I knelt on the glass with this conviction in my mind, I looked up and 'saw' the emergency squad enter the gallery in their blue uniforms, carrying bags and equipment, and looking quite competent and in charge. I was completely relaxed and felt secure in the knowledge that everything was now taken care of.

"Then I heard a puzzling question from one of the women. 'What is taking them so long? Why doesn't the emergency squad come?' I knew they were already here, so why was she so upset? Several minutes later, in a relieved voice, someone finally announced, 'Here they come now!' And in they came. The very same men in the very same blue uniforms I had just seen, carrying the very same medical equipment with the very same air of competence as before.

"With one quart of blood spilled and twenty-one stitches taken, I reviewed the experience and knew that I had been in touch with another dimension. At no time was there pain or fear, just the sure conviction I was protected and everything else was illusion."

Anna Grace was almost euphoric in her description of how smoothly and easily her recovery went and how, at the hospital, she mentally reached out to heal the attending physician of his own discomforts while he busily attended to hers. (This is often referred to as channeling, or sending "healing energy" to another through the use of mental thoughts and mental imagery.) She knew exactly what the doctor was going to say and do before he did anything, but as she verbally chronicled his forthcoming thoughts and actions to him, he was so shocked that she elected to remain silent for the remainder of her stay in the hospital. Once released, however, she further amazed the man with how fast her wounds mended, leaving hardly any scars.

Anna Grace experienced *clairvoyance* when she saw the emergency squad walk through the library doors before they actually did, and *precognition* when she knew what the attending physician would do and say in advance of his actions.

A Drug-Induced Experience

Kathleen J. Forti of Virginia Beach, Virginia, went to the dentist to have some surgical work done. She had been under the influence of laughing gas before without difficulty, but she was unprepared for the unusual experience she was about to have.

"As the dentist worked on my nerve-exposed tooth, I felt and saw in my mind's eye an intense energy pattern swirling inside my head in synchronicity with the action of the drill. I suddenly knew what the dentist's next words to me would be, how many beats of a second he would pause in between those words, and what his precise actions would be as he spoke. Then he did exactly what I had previously seen him do.

"Now, I said to myself, his assistant will say and do such and so next, and she did. A deep knowing came over me, a knowing that in this particular reality we are in essence all actors acting out our pre-assigned roles and that the play is a long-running story with continuous performances. We are destined to repeat the same sequences in our lives over and over and over again without knowing it, like Dante's *Inferno,* and we have agreed as a collective to this continuous repetition.

"No, I can change it, I can stop the pattern, I shouted in my mind, so I forced myself to say out loud the most outrageous thing to the dentist, something totally out of context. The second after I did, I knew I had spoken those very same words before in the same situation with the same response from the same dentist.

"Even my deviation was part of my scripted role. There were no surprises. Everyone already knows everyone else's lines. So why do we do it? I asked. The answer was that inhabitants in this particular reality learn from constant repetition and reinforcement of the same stimuli.

"During this experience I remember getting pictures and impressions regarding my dentist and his role on earth. I saw the sadness in his life and the problems he was facing with his children. I felt great compassion for him even though I barely knew him, nor did I know if he had children."

When I interviewed her, Kathleen said that for three days after the gas wore off, she noted several unusual aftereffects: her hearing range increased to where she could hear the thoughts people around her were thinking before they verbalized them, and she could see future events occur in her mind before they manifested in physical life. She found these aftereffects confusing, to say the least. Once back to normal, she described the whole experience as unnerving and somewhat depressing. "Here was this very real vision about predestination, but I don't know what to make of it," she complained. "I know we have choices in life, but during this vision free will had no importance."

Kathleen experienced precognition when she knew in advance what her dentist and his assistant would say and do before they did. She exhibited clairvoyance twice, once concerning her dentist's future (although her accuracy was never checked), and again during the three days of aftereffects—at which time she also displayed the *clairaudient* ability to hear other people's thoughts.

Although the exercise we have just engaged in amply illustrates futuristic awareness modes, here are some fundamental differences between the two accounts that need to be recognized:

- Both women felt they had accessed another dimension of existence during their experience—but, while the drug-induced vision *distorted* a basic truth honored throughout human history (free will), the naturally occurring incident *affirmed* a truth equally basic and valued (the power of faith and love).

- Both women had aftereffects—but, those from the laughing gas caused *instant and continuous confusion and distress*, almost a depression, then *totally disappeared* after several days; whereas those from the accident were *immediately usable and helpful*, and *continued* to enrich and enhance the individual's life *long after the event was over*.

These fundamental differences are important for us to know, especially those concerning the aftereffects. I say this because, after decades of investigating paranormal and transformative phenomena, it has been my experience that the value and meaning of subjective episodes can best be determined by an examination of the consequences, not just of the event itself. Subjective reality can be explored as readily and dependably as objective reality if one considers "before" and "after" information, keeping the event in context with the life of the experiencer and then comparing it with basic truths and historical precedents. (Refer to Appendix I for detailed material on meaningful questions to ask when seeking to evaluate subjective experiences, whether spontaneous, willfully caused, or drug induced.)

When you examine aftereffects as well as episodes, you can't help but notice that there exists yet another mode of futuristic awareness few researchers mention. This "extra" is far more dynamic and complex than the more traditional types already described, and much more mysterious. Thanks to Kathleen, we have another episode from her life to use as an example.

A Prelived Experience

When she was eighteen, Kathleen was attacked by a stranger and, with a knife to her throat, raped. Hysteria began to well up inside her as she realized she would probably die and never see her parents again. At that moment, her mind sharpened to crystal clarity, and she floated out of her body to view her attacker from a point above him. All pain and hysteria ceased, and all concerns dissolved. In this detached state, she lived a future segment of her life in great detail. In this segment, she experienced herself as an older woman telling stories to children gathered about her feet while she sat in an antique, black lacquer rocking chair, carved exquisitely in an oriental design. The paintings on the wall and each detail of the house where she lived at this future time were clear and precise, as were her thoughts, each physical movement she made, every smell and taste, conversations, emotions, plus each minute sensation of daily living.

She later forgot about the futuristic episode she had just "lived through," when she convinced herself the whole thing was a device created by her brain to ensure that she relax and submit to her attacker. Police confirmed that because of the man's history of violence, any struggle on her part would have meant death.

Five years later, Kathleen married and moved into a house her husband already owned. There she discovered the antique black lacquer rocking chair carved exquisitely in an oriental design, and the paintings, and the wallpaper, and all the details previously encountered during her near-death experience. The jolt of seeing these items surfaced a memory of having "lived" this segment of her life previously. This *future* memory prepared her for married life . . . with one exception. After six years, her marriage ended in divorce (an event she had not prelived).

Afterwards, she became interested in telling children stories and wrote *The Door to the Secret City*, a book about a child's near-death experience.[9] She then created the company Kids Want Answers, Too! and dedicated herself to teaching educators how to handle student reactions to life-and-death traumas.

As Kathleen told me this story, she was able to associate what had happened to her at eighteen with her current desire to use storytelling as a way of educating children about life's deeper issues. The more we talked, the more she also came to realize why she had been able to write such an accurate depiction of a child's near-death episode.

She explained: "I really believed back then that my experience was just an hallucination, part of the instinct for survival. But, when I walked in the door of my new husband's house and saw that rocker and all the other furnishings exactly as I had seen them before, I did a double take. This forced me to realize that what had happened to me earlier was in actuality my real future. I find it significant that regardless of the fact that the marriage ended and I moved away, I still became involved in telling children's stories—*even though the context in which I had prelived doing that had totally changed.*

"Because I divorced, I proved to myself there really is free will, that no matter what life brings we can still make a new choice from any one of many possible futures. That's why the laughing-gas vision bothered me so much at first. I knew better. I knew we could choose. My future, what I prelived at eighteen, changed because I changed. The various components still happened, literally, but not in the same manner as the original version. My choice to divorce turned everything around except the one thing I had yet to do, and even that changed in how it was carried out."

Kathleen's story about living the future in advance, every minute sensation and detail, while still engaged in present-tense time, challenges the basic tenets of how we regard existence. And her story is not unique, as you shall soon see.

I propose a new category be established under "Modes of Futuristic Awareness" to include the strange phenomenon of "living the future before it occurs." I would describe this category as follows:

Future Memory—to prelive in advance (*subjective/sensory-rich*). The ability to fully live a given event or sequence of events in subjective reality before living that same episode in objective reality. This is usually, but not always, forgotten by the individual after it happens, only to be remembered later when some "signal" triggers memory. Sensory-rich, future memory is so detailed as to include each movement, thought, smell, taste, decision, sight, and sound of regular physical living. All this is *actually lived and physically, emotionally, and sensorially experienced,* not merely watched (clairvoyance), heard (clairaudence), predicted (prophesied or forecasted), or known (precognition); and that living is so thorough, there is no way to distinguish it from everyday reality while the phenomenon is in progress.

Many of the people I have interviewed during my various research projects, especially near-death survivors, talk about the future as if it were already known and already lived, as if what they are now doing is but an afterthought, or perhaps the acting out of a previously written script. For them, *the present moment can be past tense.* This awareness, regardless of duration, is intense and may appear as precognition or clairvoyance when neither is the case. For them, it is more a process of memory than anything psychic. They actually *remember the future*, just as they remember the past, even though that memory is not based on previous "constructions" (the brain's ability to adapt actualities to accommodate whatever is precedent). They are not seers. They do not predict. They are just people who now live in a different reality system than before, where the understanding of time and space has shifted from the norm. Other researchers often misinterpret and misunderstand this phenomenon, thinking it to be something it is not.

Prepare yourself for an in-depth look at this oddity in the next chapter. But be vigilant, as clues suggesting something more might be involved are subtle.

4

The Future Memory Phenomenon

The thing that has been is that which shall be; and that which has been done is that which shall be done; and there is nothing new under the sun. Whosoever speaks and says, "Look, this is new" should know that it already has been in the ages which were before us.
—Ecclesiastes 1:9–10

Most people are familiar with the traditional types of futuristic awareness we have just discussed, and perhaps my attempt to establish yet another type might seem unnecessary or confusing at best. Yet I honestly believe the ability to remember the future—a future "lived" in advance of its physical manifestation—needs to be examined.

The people who volunteered for my study of this phenomenon came from across the United States and Canada and represent a broad spectrum of ages, race, philosophy, and employment levels. Some had attended lectures that I gave; many I chanced upon as I traveled. Others responded to announcements I had placed in various national periodicals.

Typical of the two-hundred-plus interviews that resulted from my efforts are these reports:

- A former military officer who now lives in Illinois regularly remembers the future and finds that because of this his life is infinitely more interesting and more relaxed. He hopes the phenomenon never ends.

The incidents he preexperiences usually involve conversations with people and meetings he attends.

- A woman in Washington state prelives visits to bus and plane terminals. This enables her to know in advance which travelers to look for, along with what to say to each and why. These travelers are always troubled and she sees herself as there to give them comfort and aid. "This is my job now," she related, "but I only intervene in situations where I remember having already done so. This way there are no slip-ups and I am always where I'm needed."

- A woman in Alabama finds future memory so accurate and so detailed that it even allows her to "meet" fellow shoppers before she ever gets to a given store. She is also able to preexperience the act of standing at the cash register, looking up and seeing all the other cash registers and what is rung up at each, as well as the price of many items as they are checked through.

- Another woman told this story: "I was doing the morning dishes when this rush of energy nearly lifted my head off. I suddenly experienced myself at a dinner party that night, saw who would be there, and took part in what happened and what was said. The whole thing was so real, I decided to make no plans for the evening, just to see what might happen. Sure enough, a friend called and began apologizing all over herself for being so tardy. Then she asked if I would come to her dinner party that night. I had to muffle laughter as I accepted her invitation. When I arrived at the party, it was a duplicate of what I had already experienced that morning; every conversation, every wave of a hand, repeated what I previously lived through. I'm glad I 'attended' the dinner party before it happened so I could be prepared in advance."

- A man in Indiana joked, "I don't know about this. I'll be talking with someone and suddenly I'll remember having had the conversation before, word for word, with the same person. I know what's going to be said before the other person says it. I feel lighter and somewhat giddy when that happens. Sometimes I'll just be walking down the street when it hits me, I've already done this, all of this, the cars, the people, the storefronts. It's weird. I'm still not used to it."

I discovered from these interviews that the future memory phenomenon itself generally lasts but a few seconds, or maybe a minute or two of clock time, while presaging several hours that later manifest. The phenomenon can be rather lengthy, though, and can encompass several days or even months of future activities.

Future memory, at least what I have described here, does not register in the mind as if a flash forward or a snap or a fleeting glance. This much I want to emphasize. Future memory registers in the mind as if a segment of physical reality was just experienced in its complete entirety—as if the future had unfolded in the present.

What I am about to reveal is no fantasy, nor is it the stuff of dreams. Future memory is quite real, and significantly so. The subsequent rendering is the pattern I was able to isolate from the majority of cases I studied. I believe this pattern is a universal one, and will prove helpful, not only with academic studies of the inner workings of expanded awareness, but throughout the living of our everyday lives.

The Pattern of Future Memory Episodes

Physical Sensation at Onset: There is a rush of heat coupled with a feeling of exhilaration. This rush can begin in the feet but more often begins in the lower body trunk and sweeps upward, seeming to "burst" through the top of the head. A definite "lift" is felt, whether slight or dramatic. Some experiencers report hearing a ringing sound in their ears.

Present Time-Space Relationships Freeze in Place: This instant stoppage is often accompanied by the air filling up with sparkles, as if the very molecules of the air itself were visible winking lights. Everything becomes brighter, and sense faculties heighten, yet nothing and no one moves.

Expansion: There is a definite feeling of expanding in size, becoming larger, stronger, more knowing and able. As you expand, so too does space itself, while everything else either stretches apart or fades away.

The Future Temporarily Overlays the Present: A given scenario yet to happen suddenly manifests. It is detailed and fully involved, replete with thoughts, conversations, moving, touching, accomplishments, and relating to people, places, activities, and events. Your participation in this scenario parallels actual living to such a degree that it is almost impossible to detect any difference between that which is future and that which is present. There is no awareness of causative factors (usually) or of being provided guidance, instructions, or direction. Quite the contrary. What happens is experienced as a natural component of one's life at the time it occurs.

Present Time-Space Relationships Resume Normal Activity: The scenario ends as quickly as it began. Sparkles disappear. Rightful proportions return. Animation and regular living restart as if some invisible "clock timer" were rewound. There is usually a lingering feeling of pleasure and/or disconcertedness, as if something unusual had just happened which was not expected.

Aftereffects: Invariably there is a sensation of being startled or "chilled," perhaps puzzled by what happened, with the need to seek an explanation of some kind by looking around for clues or questioning anyone nearby. The event remains vivid as long as it remains in awareness but, eventually, it is either forgotten or set aside.

The Future Event Physically Manifests: At first, there is seldom recognition, but that situation changes fairly quickly. Either a particular "signal" startles you or some key element triggers the memory of having done this before. Once memory is activated, the entire scenario returns to mind intact. This can feel spooky, funny, silly, or even "tickle," as the reality of now being but an actor or actress playing out a role long since scripted becomes an acute sensation. For many, there is a clear perception of choice and control, of knowing the script can be changed; for others, there is an acceptance or a resignation that what has been prelived will fulfill itself regardless of anyone's choice. (Unlike déjà vu, the incident clearly addresses "future" not "past.")

Resolution: Although this phenomenon can be uncomfortable the first few times, even strange or weird, it eventually becomes inexplicably appropriate and reasonable. Often the individual's life will develop a kind of flow to it or a sense of orderliness because of future memory episodes. Some people regard the phenomenon with a peculiar sense of mystery and awe; still others feel thankful, as if it were a gift from God or at least special.

Not all reports from the people I interviewed fit the pattern discussed above, but the vast majority do (including my own, for I too have experienced the phenomenon). What is listed here is the pattern for those whose experience happens while they are wide awake and actively involved in the activities of the day. There is another group whose experiences occur either during dreams, during daydreams, while thoroughly relaxed and passive, or in a state of reverie. This latter group, however, seems much in the minority, as most people are alert and busy when incidents of future memory suddenly overlay their ongoing daily routines.

Incidents usually happen infrequently, yet one woman reported ten per day during a time in her life when she was undergoing serious emotional and financial difficulties. She said the episodes gave her needed support plus the strength to keep on going. Interactions and interrelationships are the most common subject matter, rather than anything solitary or passive. In other words, future memory incidents tend to center around activity and are, in themselves, active.

Motivation to escape or avoid the harsh realities of life is noticeably absent from most experiencer attitudes, nor do any of them express any need or desire to fantasize or indulge in wishful thinking. Instead, mundane activities are those most frequently prelived, some almost boring in their ordinariness. The majority sense from this that each and every act in life has meaning and is worth doing, no matter how insignificant or seemingly inconsequential, and that somehow everything fits into a greater overall design.

I know what I have discovered about future memory sounds too bizarre to be true, but bear with me. There is more to consider.

For example, future memory may consist of an unbroken chain of minute details or it may consist of highlights in sequence. Occasionally individuals change outcomes, but even then, the original incident can reassert itself at a later date. Case in point: A woman prelived her husband having a serious car accident on his way to attend a meeting that night. This alarmed her, so she insisted on going with him and doing the driving. She thus intervened and the evening was accident free. However, several nights later, her husband had an accident and totaled the car while driving to yet another meeting, one held at a time when she could not accompany him. Her previous intervention, as it turned out, changed only the timing, not the event.

When I asked the people I contacted what they thought the purpose was for preliving the future, the most common reason given was this: prehappenings are previews; they enable you to see what's just around the corner; then they give you extra time to get ready before you get there. Some described their experiences as practice sessions or learning experiences or chances to experiment with other possibilities in living. One man described them as a window into his future.

No one I interviewed had any idea what actually caused the phenomenon to occur or why it happened to them. The majority, though, felt that the futuristic episodes enriched and enlivened their life while imparting a sense of purposefulness and meaning. They considered the phenomenon a helpful nudge toward a more sane and responsive life. Most claimed that future memory reestablished a natural rhythm to how their lives progressed, thus enabling them to possess a keen awareness that each person and every activity in life matters, regardless of who you are or what you do.

One man revealed that his parents and grandparents experienced future memory episodes, as did he and his children. In fact, several families I interviewed told me that the phenomenon was mentioned in centuries-old family records. Obviously, future memory is nothing new. Since incidents usually involve ordinary circumstances, I suspect that most people never bother to make note of it, or, if they do, what they report is probably confused with something else. Research projects generally focus on more sensational claims so media attention and grant monies can be attracted; hence, professionals seldom go out of their way for any-thing that, at least superficially, seems unimportant. I, too, would have paid the subject little heed had not the phenomenon intruded upon my own life the year after my three near-death experiences. During my initial episode, a year of advance time overlaid about ten minutes of clock time.

Opportunities for me to prelive the future have since continued, but in shorter segments covering fewer advancements of time, more often. Seldom do any of my own encounters concern dire or important events. Just as with the others I interviewed, I mostly preexperience the ordinary. This has enabled me to occasionally realign myself with something akin to what I would describe as a universal pulse or rhythm. It doesn't seem to matter what I am doing, the time of day, who else is present, or even how I feel, when such an episode occurs. What happens, happens. So far, I have not been able to cause any of them to play out as I determine.

Please do not misunderstand me. Just because my future memory episodes occur more frequently than they used to doesn't mean they hap-pen all the time. Hardly. Months may lapse in between, or a whole string of them may come, one right after another. Timing truly is random. And do not confuse moments when I intuitively "know" the future with the times when I "remember" it. These two modes of awareness differ; they differ by degree of intensity, richness of exact detail, range and depth of physical sensations involved, and probable purpose.

My story follows. By sharing what happened to me, I hope to cre-ate a situation whereby you, the reader, can examine the phenomenon of future memory in context with how it manifested in one person's life. As you read my story, note passages that illustrate what we've been talking about right along, but also pay attention to other bits and pieces that suggest that future memory might be part of or auxiliary to changes in how an individual's brain functions. I'll point out particulars at the close of each section so the phenomenon can be discussed in greater detail.

Miracles aren't always what they seem, you know.

5

Slipping between the Cracks

The future enters into us, in order to transform itself in us, long before it happens.
—Rainer Maria Rilke

I had never heard of anyone who could prelive the future in advance of its happening until I experienced the phenomenon myself.

Details are possible as I relate my story because of notes and journals I kept at the time, and because the whole thing was so remarkable it was unforgettable.

First Incident

Mid-July, 1978
Boise, Idaho

It was a Monday morning in the office I shared with two other analysts and a receptionist at Operations Improvement, a department of Idaho First National Bank, and my first day back after vacationing with an aunt and uncle who lived near Chicago. Two weeks before, I had listed my newly purchased home for sale "knowing" it would sell quickly, and it did, the very day I had returned home. My recovery from the miscarriage and severe phlebitis that had led to the health crises precipitating my dates with death the previous year was virtually complete, yet my three encounters with the near-death experience still confused me.

Perhaps it was the excitement at having returned from my first "real" vacation and having just sold my home—who knows?—but while seated at my desk, pencil in hand, something quite strange happened. Everything around me froze in place, my body included. Gushes of heat rushed through my paralyzed frame from feet to head, overwhelming me with an ecstatic thrill. As this rush continued, my desk began to fade from view while a field of brilliant sparkles descended from the ceiling and spread out to encompass the entire office.

Instantly, I was back in what I had come to call "The Void," an unusual place revealed to me during the second of my three near-death episodes. I know of no other way to describe The Void except to say it is a dimension of its own, a place where both objective and subjective time and space dissolve into the presence of all light, all darkness, all order, all chaos, all excitement, all silence, all reality, all nonreality, yet it is absolutely empty, void, except for a sensation of expectancy and a shimmer that somehow winks.

Within The Void, a swirling mass took shape, a veritable whirlpool of wave upon wave of pure energy, and I was sucked into its core like a helpless leaf. The core widened, then stopped spinning. When it did, I found myself living the life I would live until the following spring.

Briefly, this is what I prelived: Now that my house had sold, I rented an unfurnished room from a fellow analyst and slept on the floor in a sleeping bag until the time came for me to leave the state. For three weeks, I attended a party a night, saying goodbye to everyone I knew. Only what I stuffed into my small Ford Pinto went with me, as everything else I owned was either sold, given away, or stored. Four times during my trip I rested, and for four days in each place—Southern California, New Mexico, Indiana, Kentucky—before I reached northern Virginia and a temporary stay in Reston with a cousin and his family whom I hardly knew. Eventually I found a job in Washington, D.C., and an apartment in Falls Church. Another futuristic advancement in the spring enabled me to prelive my next move.

After leaving Idaho, though, I journeyed first to Escondido, California, where I attended the Death and Dying Seminar conducted by Elisabeth Kübler-Ross on August 22. I fulfilled all my childhood dreams and wishes when I meandered across the United States, including a chance to watch the sun set silver on the Pacific, then later rise golden over the Atlantic.

I explored the Carson Sinks of Nevada, meditated with the Bristlecone Pines in the Patriarch Grove near Big Pine, California, enjoyed Sea World and called upon longtime friends Diane Pike and Arleen Lorrance of "The Love Project" in San Diego, conversed with the grandfather cacti of the Arizona deserts, touched my first cotton plant in Yuma, walked across London Bridge at Lake Havasu, rode a mule train to the bottom of the Grand Canyon, ate fry bread in Albuquerque, discovered a special bear claw at Old Town, saw a miracle staircase in Santa Fe, prayed in the Air Force Academy Chapel near Colorado Springs, found a marble grave like none other in northern Kansas, hopped aboard a steam boat on the Mississippi, traced tales of Tom Sawyer and Huckleberry Finn in and around Hannibal, followed in the footsteps of Abraham Lincoln, and much, much more.

Everything I had ever vowed to do since childhood I did, as well as things unknown to me; yet each scene, each place, each stranger or friend, was important and reflected some aspect of my life. A sacred journey it truly was, for at each turn of every bend and in the depths of everyone's eyes, I met parts of my own soul looking back at me.

This scenario I both watched and lived simultaneously and in great detail, as if each physical motion was being performed in actuality, each thought, feeling, taste, touch, and smell thoroughly experienced. When the future concluded itself, The Void and everything in it disappeared as it had come, and I was where I had always been, seated at my desk, pencil still in hand. I had just experienced the life I would live until the following spring, even though I had not moved in any manner, not an inch or a twitch. As near as I could tell, about ten minutes had lapsed in clock time, although I could not be certain.

My mouth fell open and I sat there, staring, stunned. The whole incident was unnerving. I shook my head until it seemed my teeth rattled, but the presence of what had just been prelived never altered.

I could not accept it. None of it. Pulling up stakes and moving to Washington, D.C., was absurd. I was a Western woman. If I moved to a Washington, it would be to the one where Seattle was located.

As I turned to inquire if anyone else had heard or felt anything unusual or had looked my way, my body felt stiff and awkward and somewhat cool. It took several minutes before my body warmth returned and I could generate enough energy to stand and walk around. When I finally asked my questions all I received was a string of negatives. No, no one had seen or heard or felt anything out of the ordinary, and no one had looked my way. For the other people in the office, the past several minutes had been uneventful.

I was beside myself. Tingling sensations made sitting uncomfortable and thinking straight difficult at best. The pencil I had been holding fell to the desk; the work-flow analysis I had been studying blurred as my eyes filled with tears. Then the whole scenario replayed itself once more, as if to make certain I would remember each detail.

But there was more, a teaser, for the replay extended beyond the following spring to a time farther south in Virginia, when I would take a walk in the countryside: the sky was heavy with the coming of rain. A sudden turn put me face-to-face with my right and perfect partner. He was taller than I, younger, with dark curly hair and bronze skin, a conservative fellow, refined, well traveled and well educated, a lover of grand opera, unique in his devotion to the spiritual path and the reality of God. He recognized me instantly, but it took me three days to realize who he was. We married soon after, for once our hands joined so too did our hearts. A streamer of time years hence revealed twins at our side, a boy and a girl, and nearly continuous writing, public speaking, and travel.

When I could get hold of myself and quit trembling, I reached for the address book in my purse to hunt up the name of Elisabeth Kübler-Ross. Although my only contact with her had been a brief conversation at O'Hare Airport while I was visiting in Chicago, friends had given me the address and phone number of Shanti Nilaya, her ranch located near Escondido, California. (She eventually moved her center to Virginia, then later retired.)[10]

I had tried for months to attend one of her Death and Dying seminars after hearing her speak at the Mind Miraculous Symposium the previous November in Seattle, but to no avail. Enrollments were always overbooked, waiting lists lengthy, plus no policy existed whereby waiting-list names in one city were automatically added to any future seminar registration in another city. During our brief encounter at O'Hare, Elisabeth had identified me as a near-death survivor, a term that was foreign to me, and then she had described the now-famous pattern of the near-death experience. Yet right here, in my office, I had already "attended" a seminar of hers—one that would not begin until August 22, more than four weeks away. This was clearly impossible.

Reversing charges to my home phone, I dialed Shanti Nilaya. After laughing at my ridiculous attempts to explain myself, the woman who answered the telephone admonished me by saying, "Phyllis, Phyllis, Phyllis, just get here. You are already registered for the twenty-second, and I have sent you a packet. We are expecting you." I dropped the phone without concluding our conversation.

Come lunchtime, I practically flew through traffic, arriving at home in less than five minutes, a miracle in itself considering the distance. A cousin I hardly knew lived with his family in Reston, Virginia, and my uncle had given me his phone number only a few days before while I was in Chicago. A woman's voice answered the telephone after I dialed.

Again, my opening stammerings and stutterings initiated only laughter, plus a cryptic reply: "Phyllis, I know who you are. I've been expecting your call. It's Providence, you know. Providence. We've all been praying for you to be with us, and now you will. You will live with us, you know, until you can find lodgings of your own. Just ship over anything you want. We have plenty of storage room in the basement. Let us know what day you will arrive. Remember, it's Providence. You're coming here because you're supposed to." This time I managed to say "Thank you" before dropping the receiver.

Providence?

The next thing I remember is standing at the door of my boss's office, saying: "Karen, this has nothing to do with you or the bank or my job but I have decided to quit work and chase rainbows for a while."

Instantly, Karen Woods's face turned chalk white. A slight tremor caught her hand as she motioned me to sit. "Don't say another word, Phyllis. Just listen to me." She turned in her chair as if stalling for time. It was obvious my announcement had affected her. She appeared upset. In a halting voice, she began to slowly describe a vivid dream that had awakened her at four that morning.

In her dream, she had taken the bank cashier by the arm and pulled him to one side, saying: "Phyllis is leaving. She's moving away and I must replace her. Phyllis is leaving for good." Karen claimed that the dream had so startled her that she had awakened her husband to tell him about it just so she could verify that the dream had occurred.

"Phyllis, things like this do not happen to me. I do not have dreams this vivid. You walk in here, like you did in my dream, and announce that you are leaving on the very day of your promotion and pay raise, and my dream is coming true, right in front of my eyes."

Silence reigned. The impasse broke when I banged her desk with my fist and yelled, "That's not fair. You knew I was leaving before I did."

We both burst into laughter. For me, however, whatever the cause of all this, whatever its purpose, the way I had always lived my life dissolved at that moment in Karen's office.

The three weeks remaining before my departure whirled by. I rented a room from Kathy Pidgeon, another analyst with Operations Improvement, and gave away, sold, or stored everything I owned. All I kept was a sleeping bag, two suitcases of clothing, a few boxes of canned goods, files, tools, my mop, broom, and vacuum cleaner. Goodbye parties sprang up spontaneously, nightly, without word or effort on my part.

It would have been a truly glorious time had it not been for my daughter Natalie's legs. Born with a hip deformity, the steel brace used to guide the formation of missing hip joints had successfully corrected her original problem, but the procedure itself malformed her legs from the knees down. This wasn't discovered until after she had reached her teens. Numerous surgeries followed, but her continual leg pain made it clear more surgery was needed.

She had already made hospital arrangements, including postoperative care, a place to stay, and a nurse to visit her regularly. And I had cross-checked her plans to make certain no detail had been omitted. She checked into the hospital on schedule.

I couldn't handle it. I just couldn't. This wasn't part of the futuristic episode.

I sped to her side and announced that the trip was off. I would not go. Natalie looked me straight in the eyes and with a firm voice said: "Mother, you must go. You know you must. This isn't going to be the last surgery I will ever need. There will be more. Lots more. Must I lean on you forever? I'm a grown woman. If I don't take care of myself now and make my own way in life, when will I ever learn? Don't you understand? I have to do this myself. I have to. And you have to let me."

At that moment she was every inch the mother and I the child, for her vision was clear and her words rang true. I recognized this and stood silently, tears brimming.

If she was brave enough to face surgery without me, I could be brave enough to leave as planned. I could match her courage with mine. I kissed her on the lips and held her close for what seemed hours. Gently she ran her fingers back and forth through my hair. The silence was broken when she whispered, "Will you leave?"

I met her eyes and noticed how old and deep they were. How cruel it is that pain can age one so young. Nodding yes, I stood, turned my back to her, and walked away. Once in my car I sobbed uncontrollably. With hardly enough strength left to turn the key, I started the engine and drove into the darkness of a heavy night.

Although the future which came to greet me that Monday morning in my office was detailed and explicit, with place names I had never heard of nor read about, not everything I did in the coming year was part of the original scenario. And not everything I experienced that morning did I remember until after I actually began to do it.

The more graphic parts of the scenario were consciously active in my mind and I used them as "signposts" for ordering maps, charting my course, arranging rest stops, signing up for tours, and calling people in advance to say I was coming. These more graphic parts, then, provided direction. But how this odyssey came together, manifested, was truly remarkable.

As an example, when I was finally able to locate the proper phone number for reserving space with a mule train trekking to the bottom of Grand Canyon, the man on the other end of the phone said there were no spaces left. "There must be," I insisted, "for I know I am to be riding one of those mules in that particular pack train on that exact day." Just then he received another phone call and put me on hold. After a lengthy wait, he came back on the line, laughing. "Well, lady, you get your wish. One of my reservations just canceled. That's what the other call was. I can book you now."

Another example concerns the Death and Dying Seminar at Shanti Nilaya. After I arrived, I cornered Boots Martensen, the woman in charge of registrations, and asked how my name came to be on her list. She opened the registration book and showed me my name, big as life, with my address listed on Marvin Street. Few knew that address, certainly not Elisabeth or her staff, as I had only bought the house nine months before I sold it. When I pointed this out, Boots just shrugged her shoulders and laughed. "Well, Phyllis, I will say this. Your name was not in this book the day before you called. Not until that morning did I notice it here."

When I asked how that could be, Boots patted my arm reassuringly and added, "Phyllis, this is not the first time something unexplainable has happened around here. We're used to it. Pretty soon you'll get used to it, too."

According to the future I had prelived, I was to have four rest stops and each was to be four days long. The ones in California, New Mexico, and Kentucky were easily arranged with people I knew, but Indiana had me stumped. Then, a week before my departure, a most unusual cassette tape arrived by mail.

The stranger who spoke via the recording apologized for his boldness but a friend of his, he explained, had suggested several months before that he contact me and he had finally mustered up enough courage to do so. It seems the man enjoyed exchanging tapes with various pen pals and wondered if I would participate. Several times on the tape he repeated this message: "Oh, I wish you would drive into town one day soon, call us on the phone, and say you could spend some time visiting. My wife and I would be so pleased." Waves of prickly heat ran up and down my body as I grabbed for a map to check his address—Columbus, Indiana—exactly on my route. His was the missing rest stop. I telephoned immediately and asked if he meant what he had said about a visit. When he affirmed, I accepted. Rest stop number four was arranged.

But rest stop number four held a few surprises about trust and faith and the divine right of each individual to change his or her mind. The minute I drove into town, located his house, and parked my car, I suddenly "remembered" what the entire four days with him would be like: the tobacco field he would show me, the unusual buildings we would tour through, even scenes from the senior citizen's center where his wife volunteered. This memory lightened my steps as I fairly raced across the large lawn to his door. I could hardly wait to meet the couple I already "knew." But my excitement was short-lived when, after our hellos, he and his wife stiffly informed me that I could only stay overnight. This so startled me I just stood there, unable to move or think.

A loud crash, followed by the sounds of kids giggling and car doors slamming, interrupted the silence. We all rushed outside in time to glimpse the rear end of a speeding vehicle as it disappeared around the corner. I ran to my car and discovered the driver's door caved in. The couple called the police. Seeing my predicament, they proceeded to toss off the crash to joyriders looking for a thrill and said their son could help. (He happened to be the best mechanic in town.) The next day, their son shook his head. "It'll take four days, Dad, 'cause I have to order parts." As might be imagined, those four days unfolded exactly as I had known they would.

The accident made the local newspaper—because it took place at 11 p.m. on September 11 and cost exactly $111 in repair bills (covered by insurance).

Sometimes my having conscious recall of the future made a difference, but many times it did not. By that I mean, on occasion, I caused events to happen because I knew they were supposed to. But other times, events came into being regardless of outward circumstances or any manipulation on my part. If events were "on schedule" (part of the scenario, whether remembered or not), they happened. I could at times delay but not prevent their happening. These "scheduled" occurrences moved along smoothly and felt "right," as if each sequence neatly fit the next like puzzle pieces locking into place. The more I relaxed into total trust and joy, the more readily outer circumstances took care of themselves. Results were reminiscent of how air flows when uninterrupted, and that best describes how I felt—as if caught in a flow.

But if events were "unscheduled" (not part of my future scenario), there would be a marked absence of flow and no prickly remembering that thrilled, no smoothness in transition from one event to another, no meaningful relationships or purposeful interweavings betwixt and between (what some people call "coincidences"), no sense of Tightness. How I handled the "unscheduled" determined the number of "detours" I traveled before I found myself back "on course."

For instance, my "magic flying carpet" nose-dived soon after I reached Washington, D.C., when I allowed the city to overwhelm me. (You could call it culture shock.) Then my cousin informed me I would have to leave immediately because his wife was nervous about how little room they had for me. His decision felt "wrong" as if a mistake had been made. (Years later I learned that his wife had lied to him about why she wanted me to leave. Apparently I had angered her when I had refused to accept her absolutism about religious dogma.)

I left in a panic and quite by "accident" stumbled across an apartment complex in the town of Falls Church that activated "memory." Instantly, waves of prickly heat filled my body—yet I refused to rent a room there. My logical mind simply could not accept paying high rent while unemployed. I opted, instead, for cheaper housing near McLean. That lasted one week. A cockroach hungrier than I attacked me while I ate supper one evening and sent me screaming into the outside hallway. Needless to say, the apartment in Falls Church became my next address.

The McLean detour, however, drained me of all resources. I had but ten dollars left in my pocket and a tank of gas remaining in my car. That was it! Regrets heaped upon regrets as I wallowed in my admitted insanity of ever having left Idaho, and for what? To live out some crazy future already lived that I wasn't at all certain even needed living? My mood turned foul, ending all trust and joy. And, as my negativity increased, so did my problems.

A name my friend Wabun Wind had given to me returned to mind. She had once urged me to call David McKnight, a man who lived in the Shenandoah Valley and taught at the Blue Ridge Community College near Weyers Cave, Virginia. "You need to know him," she had said. (Wabun was medicine helper for the late Chippewa medicine man, Sun Bear. Since his death in 1992, she has advanced to the position of chief in The Bear Tribe Medicine Society he had founded).[11]

It took awhile to reach David by phone, but when I did, he offered to pay my gas if I would drive down for the weekend and help him celebrate Oktoberfest. The frivolity, he assured me, would restore my confidence and help me regain my courage. I went, and he was right. The next day I applied for work, was immediately hired, and began my new job. The flow reestablished itself and I was back "on course." This happened, I am convinced, because I once again became receptive to the sureness of total trust and joy and was ready and willing to risk uncertainty.

My cousin's unexpected decision and my detour of desperation were not part of the script. Yet, because of these detours, I met and connected with David McKnight. It was he who, without my knowledge, arranged for a fellow instructor at a college near Middletown, Virginia, to invite me over to present a program on the near-death experience. Two other near-death survivors attended that program, and after asking them what seemed endless questions, I was inspired to begin a quest that resulted in the book, *Coming Back to Life: The After-Effects of the Near-Death Experience.*[12]

David became my "guardian" after that, in the sense that he arranged contacts and connections for me as I searched farther afield for other near-death survivors to interview. But it wasn't until after my research was complete and *Coming Back to Life* was published, that I learned the extent to which David had quietly and without fanfare ensured that I would reach my goal. When I confronted him about this, he denied any willful intention on his part, claiming, instead, that he had only done what had felt right to him at the time. This revelation prompted me to rethink the value and the purpose behind what appears to be detours.

Second Incident

Easter Sunday, 1979
Great Falls Park, Maryland, and Arlington, Virginia

I had moved to a furnished room in a private home because the Falls Church location, although truly wondrous, had become too restrictive. On Easter morning I was seated in a canal barge being towed by two mules up the historic C & O Canal above Great Falls Park in Maryland. Storms threatened, so few were in attendance for this very first recreation of Canal history, allowing the sounds of nature full sway and giving the storytelling by the park rangers a special magic. When the barge trip concluded, I drove to nearby Georgetown and hunted what buildings and landmarks I could find that matched the park rangers' stories. My imaginary journey through the 1800s would have continued unabated had I not chanced upon an open-air concert. It was the Trinidad Steel Band, busily transforming Georgetown aristocracy into pure Calypso. I joined in.

By the time I returned to Arlington I was nearly giddy from the joyous activities of the day. A brilliant explosion of sparkles greeted me as I opened the door and walked into my room.

Sparkles?

I jolted to attention and spun around. Nothing was anywhere. Furniture and windows were gone. So, too, were walls, floor, and ceiling. I had somehow stepped from one dimension into another when I had innocently entered the room. On one side of the still visible door, it was midafternoon and broad daylight. On the other side, where I was, The Void awaited, not illusory, but as a physically manifest realm.

Spring had come and, as my first session in preliving the future had revealed, another advancement of time was now due.

So, within this sparkly brilliance, I prelived the next phase of my life, a phase in which I would move to Roanoke: *Friends from the District and my son assisted with the move, and two special people invited me to live with them. They were a retired couple, short of height, stocky, active, and gifted with a great sense of humor. Once again, Diane Pike and Arleen Lorrance shared time and space with me as we raised our voices to participate in a class they were giving on the healing properties of sound. I returned to Idaho, rented a truck, and hauled the last of my belongings to Virginia. As part of my new job, I engaged in writing pages upon pages of material, nearly endless writing. Then, while taking a walk along country roads as a downpour threatened, I turned around and suddenly faced my right and perfect partner. We married six weeks later.*

Brief but specific, the scenario ended and the sparkles disappeared. The room returned to normal with me standing in the middle of it. Sun shone through the windows. I dropped to my knees and yelled, "No one in their right mind will believe any of this, and besides, what's a Roanoke?"

On the nightstand was a stack of mail I had forgotten to open. When I made a grab for it, the stack flew apart and paper scattered everywhere. One by one, I gathered each piece until I came to a green brochure. A photo of Diane Pike and Arleen Lorrance caught my eye.[13] Half shaking, I ripped open the flier and read that in May they were going to present a workshop in Roanoke, Virginia, on how to heal the inner self through the art of toning (where you create healing sound by using the human voice in a certain manner). When I had visited with them the year before, they had made no mention of this. And Roanoke? Obviously it was a city.

Immediately, I did three things: wrote a letter to the Roanoke Chamber of Commerce requesting a map and information about the city, made out a check and included it with my registration for the workshop, and constructed a poster. The finished poster consisted of a photograph of myself, a brief resume, and the fact that through "guidance" Roanoke had been suggested as a place where I might live. I hand-lettered on the poster the question, "Does anyone have a room to rent or know where I could live should I move to Roanoke?" I then addressed the check, registration, and poster to the Unity Church of Roanoke Valley, where Diane and Arleen would be holding their May workshop, and dropped the parcel in the mail.

Responses came quickly. The Chamber of Commerce sent an intriguing packet describing the city as located in the southwestern part of the state and ringed by the Blue Ridge Mountains. It's nickname was Star

City. With map open wide, I spontaneously pointed to a street in the northwestern sector and shouted, "That's where he lives, the man I am going to marry."

The next day, I received a letter from a retired couple by the name of Don and Neddy Repp. They had just moved to the Roanoke area from Ohio and had a vacant bedroom in their large home outside of town. Their letter read: "The Church put your poster on the bulletin board and we both recognized you right away. We know you. You're part of our family and our extra bedroom is yours if you want it. We'd like very much if you would live with us."

That did it! By the end of the week I had resigned my position in the District and told my coworkers it was time for me to return to the land of pastures and children and corrals and barns, where I could write a book. It was now or never.

It took months before all arrangements with my son and several friends could be made, but move I did. After several false starts looking for work in the Roanoke area, I hired on as a switchboard trainer with a fast-growing telephone company that specialized in computerized, high-tech installations in large hotels, motels, and businesses. The book I had planned to write refused to be written, so I shelved it. But, just as I had prelived that Easter Sunday morning, I did indeed become involved in almost continuous, nonstop writing on my new job—manuals. And how that came about was a surprise.

One day, although I knew absolutely nothing about computers and even less about telephone switching systems, I challenged my boss about a particular manual all trainers had to use.

"Let me rewrite it," I begged. "No," he stated flatly, "only our engineers do that."

"But it's terrible," I moaned; then I proceeded to nag him with a barrage of complaints. More to shut me up than as a statement of faith in any ability I might have, he conceded. He arranged for me to have free rein in a business complex where a system like the one described in the manual had just been installed. I could push the equipment, experiment, do whatever necessary to obtain the information I needed to produce the manual I wanted.

For reasons beyond explanation, my mind could visualize and trace the entire switching system and how it worked from the inside out, and do it perfectly. Not only did I rewrite that manual, I wound up rewriting most of their other manuals, as well. Even the engineers were impressed.

My job title changed to telephone systems analyst and systems troubleshooter, while no one was more amazed than I at this turn of events. And, while familiarizing myself with huge cabinets of equipment and machinery, I fell madly in love with every single switching unit I encountered. I would gear my time so as to spend hours alone with the machinery, fondling each piece I could safely touch, drinking in the power that surged through the circuitry. I "knew" this stuff as surely as if I had been its inventor, yet, not once in my entire life had I ever exhibited any inclination for mechanics or computers or technological inquiry.

Yes, I did return to Idaho, rent a truck, and haul back to Roanoke the last of my things. But, because my job necessitated constant travel, I had not returned to the Unity Church for some time.

One weekend, though, when I was in town, Don and Neddy asked me if I'd like to attend a special Friday night program at the church. The topic was Zen meditation, a particular favorite of mine. I attended more to support the program than with any curiosity about the subject, for I could hardly believe a church in Roanoke would discuss Zen. Don and Neddy were excited when they saw who else had come—a particular man they had wanted me to meet. When they insisted, I greeted him, but only to please them. My search for my right and perfect partner had led nowhere; hence any idea of male companionship had ceased to be of importance to me.

The next day, a day heavy with the coming of rain, I was outside hiking along a country road near the Repp's house when I turned around and was suddenly face-to-face with the same man from the night before. He was taller than I, younger, with the dark curly hair and the bronze skin of an African American. We walked the rest of the way together, hands joined, chattering like long-lost buddies.

He was a conservative fellow, rather refined, well traveled and well educated, a lover of grand opera, and unique in his devotion to the spiritual path and the reality of God. He claimed to have "recognized" me at our first hello, yet nothing registered in my own mind until three days afterward when my heart finally reminded me who he was. We were married six weeks later at the Unity Church of Roanoke Valley. Although I was unable to have more children (hence no twins), it is interesting to note that Terry himself is a twin, and on more than one occasion he has commented on how much I seem the other half of him, as if his twin soul.

When I pointed one day to that spot on the map where I thought my right and perfect partner lived, I asked Terry to tell me how close my finger was to where he was living before we met. "You're a block and a half off," he replied, "but when you first located that spot on the map I wasn't there. I didn't move to Roanoke from Louisville, Kentucky, until June, several months later. You pointed to *where I would live*, not to where I was living at the time."

6

Learning to Remember

The moment one gives close attention to anything, even a blade of grass, it becomes a mysterious, awesome, indescribably magnificent world in itself.
—Henry Miller

More fairy tale than conventional reality, the story I have just shared with you was as much unnerving for me as it was incredible. Yet there is no denying truth, however eccentric or exotic, and true it was. Evidence stands. No one's belief or disbelief, including my own, changes anything.

The whole phenomenon of future memory seemed like a miracle to me the first time it happened. It felt as if I were caught up in some kind of supernatural flow, perhaps a cosmic breath, whisking me along pathways as magical as they were mysterious. The romance of it all far exceeded any fantasy. After the second one occurred, however, and especially after Terry said that the spot I had pointed to on the map was where he would live, not where he was that spring, well, my senses went on full alert.

Yes, ever since my near-death experiences, things stranger than I was used to had been happening to me, and with increasing frequency. Yet, this ability to prelive the future seemed of utmost importance, as if it were a signal of some kind. When I began to investigate the phenomenon in earnest, I kept pondering why so many different people could remember the future as well as the past. And, I asked myself, what for?

Then I met James Van Avery.

A senior electronics design specialist for an aerospace company in Seattle, Washington, Van Avery came to my attention in June of 1989, when I presented a paper covering my research on the aftereffects of the near-death experience at the International Conference on Paranormal Research, held in Fort Collins, Colorado. He also presented; his paper was entitled "Remote Viewing Using Future Memory Technique."[14] I hurried to his room, arriving as he boldly exclaimed: "We can live the future in advance by deciding to, then practice and practice, hone the skill, until it becomes natural to us." I couldn't believe my ears. Naturally, we had dinner together after his talk.

As we visited, James Van Avery reminisced of a time when, at the age of ten, he had chanced upon a roulette concession at his father's company picnic. Instantly, he had known exactly what number to bet and saw himself winning. Later he placed that bet and won as he knew he would. "Ever since then I have been trying to duplicate that event," he explained. "Not just the winning, but the knowing in advance."

After a decade of experimentation, he finally developed a method of how to remember the future, and in so doing, he fulfilled his goal. Even though the reality of future memory violates present understandings of time and space, he was enthusiastic in his conviction that anyone could learn the technique and develop the ability to perceive the future, as well as select objects or events in advance of manifestation (thus manipulating the future). His list of successes was pages long. "Practice makes perfect," he asserted. "You can remember more when you keep remembering more. Persistence is the secret. Make a game of it, like 'What's in the box? What will be in my hand?' Then imagine what it is like to open the box, lift the object, and put it in your hand. You can perceive not only what's going to be in your hand before anything is there, but also what the object was before it ever got in the box or even before it was ever purchased or made. You can see what happens before it happens. Practice. Write everything down. Although sporadic at first, accuracy percentages will increase in ratio to how often you practice."

As we parted company, Van Avery noted: "The only completely convincing paranormal event is a personal experience. No amount of documented reading material or controlled demonstrations will change a personal belief system. My technique is designed to allow you to find awareness the only possible way—*on your own!*"

He's right. Thus, with his permission, and in recognition of the many people who have been able to replicate his claim using the method he developed, I offer the following:

James Van Avery's Future Memory Technique

Developing future memory is a very exciting process, which requires persistence and patience. Expect results that will give you an incomparable reward. Do all of the exercises in order and don't skip ahead!

Number One—Getting Started

Find a quiet resting point during your daily routine. A prime time is when you have completed a chore or effort you feel good about finishing or at the completion of a challenge that you have won. Whenever this feeling is sensed, use it to your advantage in the exercises to follow. It will assist in achieving good results.

Standing or walking is all right, but in the beginning, a sitting position is best. Gaze at a scene that has many items such as flowers, vase, table, rug, wall picture, telephone, your typically cluttered room will do just fine. When you feel you can remember a lot of items in the scene, close your eyes and start visualizing the scene. You will notice that many of the items that first drew your attention are easy to recall, at least you think so. Now open your eyes and compare. Do not spend more than three seconds with your eyes closed. Keep your mind focused on your recall scene. Change the eye-closing time to fit your recall pattern needs, sometimes a short time closed, like one second, then three seconds open. Be creative during this effort. Improving your memory is an art form, not a science.

Feel comfortable while performing this activity. Always say in your mind, "How did it look when I just saw it?" Visualize what the scene had in it during the last gaze. Do this until you feel very confident that you can maintain complicated and accurate scenes in your memory recall bank. Play games with the scene and look for details that at first seemed insignificant. Scrutinize everything, like grains in woodwork, textures in carpet, small patterns in wallpaper, even dust on furniture. Become one with the scene. Bounce back and forth from memory to vision, eyes open then closed, open then closed—remember!

After doing this for a while and feeling very comfortable with it, mentally step outside of this process and observe what you are doing. Meanwhile, continue to remember more accurate details of the scene. Feel the parts of your mental process that are being activated during this procedure. Notice how simple the process is becoming and how sure you are of yourself—confidence! You may even feel you are falling into a trance, or entering a state of not knowing why you are doing this. Simply forget that you are even trying to remember. Think of yourself becoming part of the scene, which should be easy now that you know so much about it. If this begins happening, you are making good progress.

After a while, you will notice you are getting stuck on insignificant details and focusing on smaller and smaller portions of the scene. It will become difficult to see the whole scene because your concentration will shift to details. Periodically, put the details together to create the whole scene more accurately. Do not concentrate on any one scene for more than a minute. Keep speeding up this process the more comfortable you feel with it. Get in the habit of doing this with many scenes during the day when you find the time to relax and get into the flow.

This exercise is very important in the beginning to create a basic feeling for future memory visualizations. Keep repeating it until it becomes second nature. When you find yourself doing this without initiating it consciously, advance to the next exercise.

Number Two—Imagination Is Real

After you feel you have mastered exercise number one and you are ready for some new material, slowly start injecting small amounts of imagination into the scenes. For instance, "I wonder what the vase is made of?" "How thick is the paint on the wood?" "What does the inside of the flower look like?" "How does the grain of the wood look inside the board?"

Use your imagination and visualize what things look like that have other objects in front of them. For instance, a book located behind a lamp. Items you cannot visually see, but imagine should look a certain way because they are symmetrical.

Do this with integrity such that you feel you are probably correct about your hunches. Don't worry about comparing your imagined thoughts with what is actually in the scene. You are building confidence in your imagination that you have always taken for granted and assumed is not real.

Number Three—Mind Mirroring Reality

When you feel comfortable with the thought processes described in exercises one and two, spend a fair amount of time reflecting on how your mind works. Do this until you feel you have fully explored the activity of your recall process. Sometimes during all of this, thoughts appear to you that are not directly related to the scene being studied. This is known as clutter. Don't worry; you can always filter out unrelated information later upon reexamination of your past thoughts. However, strive to subdue clutter and keep it to a minimum. The important thing here is to be aware of your own personal way of keeping track of information in your head. You will need to be very accurate in identifying mind wandering and clutter during future memory targeting.

Number Four—Record Keeping

I cannot stress enough, *keep records*! This is why I am taking the space to make it a separate exercise. You are probably saying, "I know what's going on; why do I need to write anything down?"

Here's why. If you do all of these exercises and stick with the technique, you are going to change a lot. It will be gradual at first and somewhat unobtrusive as the days go by. When you analyze your records, you will see your progress and notice changing patterns in your future awareness. When those really exciting scenes come into focus, you will be in the habit of keeping them like precious photographs you took of your first child or honeymoon. Some of the scenes will give you goose bumps and adrenaline surges, indicating that you are certain you have controlled your awareness to actually peer into the future.

If you can draw, use this talent to your advantage. If you can't, do the best you can to learn. Sketching out your scenes is the best way of recording images. Words are good, but drawings are a whole lot better. When you target your future scenes, sketch them out before you go for the proof. Use lots of words and phrases to enhance the details on your sketches. Grade yourself on each sketch. Always give yourself partial credit for any accuracy. Parts of a sketch may be totally wrong, while other parts may be perfectly correct.

By studying your results, you will see what you are doing right and what you are doing wrong. This will be your best guide to perfection. Persistence will pay off, believe me.

Having records to refer to will help to show others in your life how you are doing. It will give you increased confidence when you are called on to demonstrate your skills. You will be able to help others by showing them your personal ideas that worked best.

Number Five—Patterns and Shapes

When perceiving scenes of future linear time events, always use the statement, "What will it look like when I see it in the next few minutes?" Never say, "What is it?"

Do not try to close, or guess exactly what an object or scene will be during this phase of your development. Never say what the object or scene actually is, only what it looks *like*. This is very important. Closing will come at a later time, but for now it will cause more harm than good.

This is one of the best exercises I have found to train yourself to actually project your consciousness slightly ahead in time and build confidence. It is good because you can do it all by yourself. No need for a partner who might give you negative feedback at this infant stage of learning. It will start your awareness moving in the right direction. Always trust your imagination. Don't forget, it's real!

Gather a stack of about twenty colorful magazines and a sketch pad. Say out loud to yourself, "What will the magazine page I will see in the next few minutes look *like*?" Be very specific about your question. Now draw general shapes of what you think will be on the page when you see it. Give yourself about two minutes to draw some fleeting images. Don't try to dream up scenes. Let the pieces come to you. It will appear that you are getting a momentary glance at an unfamiliar object. Get in the habit of working quickly. Your imagination will turn to fantasy given too much time.

After you feel that you have given it your best shot, randomly grab any magazine, open it to any page, and stare at that page for about a minute. Study it similar to the way you learned to observe scenes in exercise one. Now look at your sketch. Notice any similarities? If yes, give yourself a score. If not, look harder. Bounce your memory back to when you were first drawing your sketch. Do you remember things you didn't include that do look like the picture you chose? Almost immediately you should notice some similarities. Give yourself a score right or wrong. Save all these sketches in a scrapbook. You will soon notice them getting better and better.

Some may be so good that you will be tempted to show others the amazing coincidence.

During this phase of your development, you will probably relate these similarities in your sketches with the selected photographs as coincidental, no matter how many times it happens. This will be all right for now, since you are just starting to become more aware.

There is something very important to watch out for in this exercise. It is known as overlay. If you attempt this exercise too many times in a row, you will have trouble clearing the previous scene from your consciousness. Pieces of past images start appearing in your new future memory scenes. With practice you will come to identify when this is happening. Be aware of it always. It can be somewhat difficult to identify. If it becomes too difficult to cancel out or identify, you are trying too hard. Think about something else for a while until your mind clears.

Another indication you are trying too hard is when you have drawn a beautiful sketch of shapes and objects and the page selected turns out to be all newsprint. Obviously, you were way off base. When this happens, you are projecting and not allowing the scene to come to you.

If you get tired of magazine pictures, you can use real-life scenes. They work exactly the same and are much more interesting.

Your accuracy will be the same; however, you cannot compare and learn from your results. Comparison of magazine pictures and sketches is the best training device.

As you are walking down a hallway, visualize what it will look like when you turn the corner. Will there be boxes in the way? Are there any people in the hall? What will they look like? What are they doing, wearing? Ask yourself a lot of questions. Be creative. Make a game out of this effort so that you have fun. Keep your goals light. Don't take yourself too seriously. Don't push too hard as to stress the naturalness of the imagination process.

All through this exercise you should be looking for shapes and forms of light and dark. Contrast is very important. At this time they will appear to be in black and white. However, always strive to bring color into your visualizations. Images will appear blurry and smeary without sharp edges. You will notice some details periodically, though. When this starts happening regularly, you are ready for the next exercise.

Number Six—Identifying Details

At this point in your development, you should have noticed a lot of coincidental circumstances. Not just in your future memory sketches and real-life scene visualizations, but you should now be more consciously aware of spontaneous happenings in your daily life. For instance, "I knew the phone would ring. I knew it would be you on the phone. I knew that car was going to swerve into my lane."

Coincidences are appearing more frequently now, but you are still not convinced anything out of the ordinary has changed. It simply seems that synchronicity has become more noticeable. This is because you have not yet been able to control and identify real, accurate details.

Remember in exercise one when you were practicing viewing and remembering details of scenes. The overall scene started to fade away while details got bigger and bigger. It appeared that the details became the whole scene. You had to pull your vision back and patch the details together to form the big picture. When you did this, the entire scene became more accurate and precise.

Expand your newly learned talents in this exercise by modeling your consciousness after the analogy of a fisherman casting a fly on the water.

Whip your perception in and out of the scene. All the while, mentally situate yourself outside of this action to observe the results. It's really not as hard as it sounds. Actually, it's a very natural process once you get in the habit. This is a way of not getting stuck on any one detail. Keep moving; then piece the details together later. The reason why details seem so hard to grasp and retain is because you are working with the stuff that makes up dreams.

Number Seven—Future Memory

A very good analogy to explain how future memory works can be best illustrated using the following hypnosis illusion.

Using the regression of a client, a hypnotist is often trying to get to a scene that has been blocked from memory, such as rape or child abuse. This scene has caused a lot of problems in the client's subconscious and cannot be erased or repaired without conscious recall. If the hypnotist works from present time back to the problem scene, he runs into a brick wall. The client simply will not allow this part of his memory to be accessed. One surefire way around this is to regress the client back to a period in time before the scene took place. Coming forward in time from this point allows the client to approach the scene somehow without the subconscious blocks.

This illusion can also be used to go ahead in time, hence the term "future memory." When advancing to a future event, visualize yourself further ahead than the target time. Now use your memory to remember back to the time you wish to observe.

Picture a room that is familiar to you, that you can enter in the near future. Strongly imagine yourself standing in that room. Pretend you are already there. Now, using your memory, start remembering how the furniture is placed. Look for details of the room, using the fly-casting fisherman technique explained in the last exercise. Draw all this information on your sketch pad. When you are finished, physically go to the room and check your results. Give yourself a score and keep all records.

It will appear that the more thoroughly you examine the room, the better the results in your sketches. You will discover that completing the cycle of confirming your results somehow allows you to perceive more accurate images.

Number Eight—Transformation

If you have faithfully completed all the exercises so far, you are a survivor. *Congratulations*!

The next phase could be the most difficult of all. At this point, you must make a drastic change in your belief system that what you are doing is real, actually happening, and can be controlled. The thousand or so records you have been keeping will be helpful to make the transformation, but once again, beliefs do not rely on logic for justification.

By now, you should be convinced that your mind, memory, imagination, or whatever you want to call it, can and has violated the barriers of linear time and conventional space. Once you accept this and alter your belief system, something wonderful will happen. All that struggling you have been going through up until now will end. Being conscious of future events will seem almost as easy as remembering what you ate for lunch yesterday. You will actually wonder why visualizing future events was so difficult. It has been tedious, simply because you didn't believe it was possible.

Number Nine—Problems and Solutions

In the beginning it seemed easier to associate with emotional feelings such as loved ones or changes in your personal life. In fact, it was probably an emotional experience that opened you up to the possibility of being conscious of future events. It is true that having a strong emotional reason for needing to know a future event makes the view more clear and vivid, but with practice, it will be easy to associate with anything you wish. Feelings, objects, places will make no difference. Just be aware that emotion is not always necessary but it should be involved, or you will develop a "so what?" attitude and lose interest. I have found it is very rewarding and emotional just to be correct. I hope you develop this same positive feedback.

The following thought has helped me to relate to real-life scenes. Since I am going to physically see the object or scene in the next few minutes, it does not logically seem any harder than drawing the magazine pictures as in exercise five.

In the beginning, I assumed that the longer the time gap between sketching and observing the object, the harder it would be to get correct results. This turned out, like most other assumptions, to be something I chose to believe and was incorrect.

When checking your results from a room scene that you will walk into, always examine your results from the exact spot you had in your imagination reality to get the correct perspective. What you see is what you get! Chairs look very different from different perspectives. Do not imagine yourself on the ceiling or up above eye level, because you can't physically get there to check out that perspective.

Don't let wrong results discourage your progress. Simply use them as a means of learning to be more correct.

From personal experiments I have acquired many helpful hints to eliminate problems, but space constraints do not allow a detailed accounting. When in doubt, simply rely on plain old common sense.

Conclusion

I have attained a high degree of accuracy using the Future Memory Technique previously described. All that was really required of me to do it was a change of focus and persistence. No rituals or magic, just accurate and honest documentation, then careful examination of the results to steer the learning process in the right direction. Being successful has changed my life, knowing that I can be aware of future events. I am very proud of this achievement and enjoy sharing what I have found with others.

James Van Avery's technique for learning how to remember the future is remarkably simple. The key to its success is commitment, your commitment to persist should you decide to do it.

Meeting Van Avery put a whole new slant on my investigation of future memory. His accomplishments forced me to rethink the phenomenon, not as some unique or incredulous anomaly, but as a ready sign that an individual's brain structure and brain capacity can indeed *change*.

Part II

The Innerworkings of Creation and Consciousness

7

Deeper into the Labyrinth's Depths

*The Universe is immense and gorgeous and magnificent. I salute it.
Every speck, every little fly on the window salutes the Universe. Every
leaf has its meaning. I think the Universe is expanding—it is experiencing and accomplishing. And we have the opportunity to add to its glow.*
—Helen Nearing

We are not stuck with the intelligence potential we inherited, nor with what has developed within us during the years we have lived. A brain shift, whether triggered by a spontaneous event (turbulent method) or occurring in conjunction with spiritual disciplines (tranquil method), is within the reach of each and every person. Brain shifts can even be caused. Slight alterations in consciousness are not what I am referring to here, but change, perhaps even a significant restructuring of brain function and chemistry.

People who undergo such shifts display a childlike innocence and simplicity afterward, along with a marked increase in intelligence. Most report future memory episodes. Compare that with what is known from the wealth of research done on childhood development.

Children do not have a natural sense of time and space. They learn this primarily between the ages of three and five by projecting into the future and by intuitively engaging with futuristic ideas, images, feelings, and sensations. It is important to note, however, that awareness of the future *does not appear as "future"* to children. To them, it is but another aspect of "now" (that which is immediate), and remains so until they are able to establish the validity of continuous scenery and connected wholes. Once they do this, they gain the perspective and continuity needed to adapt to ever-changing environments and the meaningfulness of life.

Research with the young also reveals that at each major stage of their development, there is a spurt in brain growth. It is then that the natural process of brain neurons (nerve cells) combining to create field effects (potentiality) suddenly accelerates. How enriching the environment is and the models he or she associates with during these spurts determine the extent to which intelligence and creativity are enhanced.

During the research of near-death states and the experiments I conducted in the sixties to explore the validity of spiritual transformations, I kept noticing correlations between what children go through when they discover life potentials and what adults go through when they *rediscover* the same thing. The farther afield I searched for answers, the more apparent it became that I needed to learn how to ask different questions—questions that invariably led me back to the brain and multiple aspects of mind.

I began to ask questions such as: What if a brain shift is the process nature uses to ensure the continued adaptation and evolution of the human species?

What if the trait of being able to embrace the future in the present is actually an indication that both perspective and continuity can be maintained in ratio to each stage of brain growth that follows a brain shift, so the impact on the individual's life can be eased?

What if the "otherworld journeys" we experience during a brain shift are really our brain's way of sifting through and reassessing the early models and patterning it originally adopted and updated during the span of its existence, however short or long that time span?

What if transformations of consciousness, no matter what type, are but "housecleaning" mechanisms that set the stage in the brain for higher and greater levels of mind to emerge?

I find it fascinating that between the ages of three and five is when most childhood cases of alien abductions and alien sightings are reported to occur; when most early childhood cases of the near-death phenomenon happen (especially if that close brush with death involves high fever); and when kids commonly report paranormal activities like out-of-body experiences, flying dreams, disembodied voices, spirit visitations, and heightened intuition.

Children between the ages of three and five inhabit futuristic scenarios on a regular basis and identify with and assert their ego in creative and inventive ways. Memory formation piles images atop images as the young construct their version of the world around them, building societal stereotypes that help them make sense of the larger, universal patterns (archetypes) they seem to have been born with.

Research on childhood development suggests that our temporal lobes are the pattern holders, that place in our brain where basic imprinting accommodates an ever-changing array of imagery. It seems to be where our original "blueprints" of shape and form are stored.

From the research I did on near-death states, I came to recognize various correlations between archetypes, stereotypes, and personalizations as being linked to both the ages of children who had near-death episodes and the patterns of change most adult experiencers grappled with afterward. It is my hope that, as I share these observations, they will help to shed more light on the natural process of brain development and memory formation *in tandem* with what forces may be at play during otherworld journeys.

Obviously, there is more involved in otherworld journeys than mere concerns about past-present-future and what some might term "paranormal." Rather, there seems to be a larger dynamic at work that arranges and rearranges reality according to what is needed at any given point in our development and within the universe at large . . . while restructuring our brains and reawakening our hearts.

As we interpret our otherworld journeys, we reinterpret life's meaning and purpose.

As we explore how consciousness can change, we reexplore creation itself and the wonder of the universe (see "Changing Patterns in Mental Models" chart.)

As we find order afterward, we reorder our world and our place in it.

CHANGING PATTERNS IN MENTAL MODELS

ARCHETYPES	STEREOTYPES	INDIVIDUATIONS
Broad, larger-than-life shapes, patterns, themes of universal import; timeless, impersonal, detached (i.e., angels at a distance holding out their hands in a gesture of caring as they encircle a spiral to higher realms).	Societal engagements, structures, activities geared to environmental and societal expectations (i.e., angels holding each hand as they walk with you and talk about your life and the way you lived it, an awareness of rewards and responsibilities).	Individual assertions, relates to personal curiosities and the need to identify one's rightful place in the the overall scheme of things (i.e., either challenge or coparticipate with angels while exploring new options and new choices in life).
Time Awareness: Now; based on immediacy of simultaneity.	Time Awareness: Short-term; based on shared consensus.	Time Awareness: Long-term; based on priorities and values.
AGE LINKS TO THE KIND OF IMAGERY FOUND IN CHILDREN'S NEAR-DEATH EPISODES		
Infancy	Beginning 3 to 5 yrs	Beginning with puberty
REACCULTURATION PROCESS MOST ADULTS GO THROUGH AFTER NEAR-DEATH EPISODE		
Phase One (lasts about three years): More cognizant of archetypes, universal themes and issues, greater realities, other worlds; impersonal, detached from ego identity/personality	Phase Two (lasts about four years): Concerned with societal interaction and inter-relationships, service and healing oriented; realigns or alters stereotypical roles, community oriented	Phase Three (usually after seventh year): Aware of self-worth, practicality and discernment emphasized; tends toward self-governance and self-responsibility and spiritual development

The deeper we descend into the depths of the labyrinth this book encompasses, the more surely we plunge into a vast cauldron of paradoxes. Nothing we find will make sense unless we are willing to confront the actuality of what we think is real and consider alternate viewpoints.

The novelist and short story writer F. Scott Fitzgerald wrote: "The test of a first-rate intelligence is the ability to hold two opposed ideas in the mind at the same time, and still retain the ability to function." Current discoveries in physics have validated his statement—that multiple realities do indeed exist, and that neither singular solutions nor either/or decisions need apply.

Clearly, infinity is not bound by the concepts of individuals looking at it.

Now that the groundwork for a basic understanding of the future memory phenomenon and its probable purpose has been laid in part one, we are ready to fit that understanding into a larger framework, to give it context. To accomplish this feat, we will take one giant stretch after another throughout the entirety of part two, as we touch on material both controversial and surprising. If our pace seems to slow at first, don't be fooled by that. We begin at the beginning with consciousness and flow states, the brain/mind complex, and tricks of perception. But, from there, we play leapfrog with new frontiers of thought and experience.

Ah, but that's how it is when one traverses a labyrinth. Just when you think you're making headway, the whole thing folds in on itself and you wind up having to encircle your goal rather than proceeding straightaway.

No complaint intended.

After all, mystery isn't fun if it isn't . . . mysterious.

8

Connections within the Flow

All are but parts of one stupendous whole!
Whose body Nature is, and God the soul.
—Alexander Pope

Take note: Future memory happens most frequently to people who are wide awake and actively engaged in life routines, people who suddenly and without warning experience the cessation of time, space, and reality. For them, everything freezes in place . . . *except consciousness*.

Once the earthplane and the laws that govern it cease during a future memory episode, the experienced consciousness is freed of any constraints or interference from brain activity. His or her consciousness, thus released from the necessity to think, *expands automatically*. Generally, there is an accompanying "feel" to this expansion that is predictable in how it is physically felt. The majority describe this feeling as a rush of energy, often upward in direction, either toward the head or out past the head. If the rush is not felt, then something akin to a prickling sensation or exhilaration or "thrill" is experienced. When this happens, consciousness seems to align itself or connect within a greater stream of intelligence, a "flow."

What do I mean by "flow"?

It has been my experience that there is an internal state of flow and an external condition of flow, plus the puzzle of synchronicity.

All three seem to lead beyond the threshold of rational mind to the existence of yet another type of flow—the way one's entire life can begin to arrange itself by itself as if fitting into some larger plan.

To understand what I'm edging toward, let's take a look at the two main aspects of flow and what they may indicate about future memory:

Internal State: Complete absorption in a subjective/mental flow state either momentarily (by blanking out), or for sustained periods of time (by cultivating the enhancement of flow or through what is commonly known as meditation).

External Condition: Living one's life in an objective/physical condition of flow, where unrelated events come together by themselves in ways both curious and meaningful—with a peculiar ease of timing.

Flow states are influenced by the receptivity we have to them and by how easily we can relax and "allow" the flow to happen.

Internal State of Flow

Mihaly Csikszentmihalyi, a psychologist at the University of Chicago, is the leading researcher today in this internal state of mind, and his findings are pertinent. He defines an internal flow state as being so absorbed in what you are doing that time and space cease and a euphoric feeling of complete clarity and sense of purpose takes over. Being in this state of mind he refers to as "going with the flow."

People lose a sense of self in this state. One becomes both actor and observer, irrelevant stimuli are shut out, time and space distort, and there comes a knowing.

Csikszentmihalyi reports that an individual in a flow state steps beyond the limitations of self and encounters unlimited possibilities. Brain activity measurements change. Unlike concentration, which increases cortex action, flow states decrease cortical activity. Speculates Csikszentmihalyi in his book *Flow: The Psychology of Optimal Experience*,

"We get into the flow not by exerting more effort, but rather by screening out distortions. That would mean flow resembles Oriental meditation practices—the notion of learning to stop the world."[15]

Flow also resembles that moment when the creative urge leads an individual beyond the bias of thought, as well as that state of mind many future memory experiencers enter into as the phenomenon begins.

Of interest is the fact that children, eager to match what they have learned from new challenges in their life, have flow states all the time. Yet, research findings from numerous scientific sources establish that anyone at any age can have them. This means that not only can such a state of consciousness become a part of everyday living, it can become a desirable and positive adjunct to life skills. Frequency of occurrence distinguishes flow states from the transcendental peak experiences investigated by Abraham Maslow in the early 1960s. Maslow described a similar inner euphoria, but he maintained that it could only be experienced a few times in any given lifetime. Csikszentmihalyi, on the other hand, finds that "if you cultivate the flow, you can have the experience several times a day."

Do not confuse any of this with daydreaming, because flow states are really periods of blanking out, of reaching zero, and then simply flowing with the feeling of being there. Flow is letting go—into nothing. You do not actually go anywhere, yet you feel as if you have been everywhere—as if all that you know or ever could know converges with all that is. Nothing happens, yet everything happens. You emerge from a flow state refreshed, inspired, and invigorated—almost as if you had been asleep—yet you know more than you did before, and you have no idea why or where the information came from. Flow has nothing to do with drugs or electrical pulsations massaging the brain. It is a natural state that is arrived at naturally.

Flow opens wide a doorway to unlimited realms of wisdom and knowledge and endless possibility. And it is a major factor in explaining how multiple realities can be accessed by anyone, at any time, anywhere. Knowing this helps us to conceptualize how future memory might occur. That's because flow and the phenomenon of preliving the future have a lot in common.

External Condition of Flow and the Puzzle of Synchronicity

Flow can refer to more than just a state of mind, however. It can also refer to a unique arrangement or series of meaningfully connected events in our outer physical world. To put it another way, experiences can flow, opportunities can flow, interactions can flow, activities can flow, our life can seem as if caught up in an external condition of flow.

Csikszentmihalyi neatly mapped the inner workings of flow, and how this state of consciousness could be better utilized for personal benefit. The outer condition of flow, as concerns physical manifestations that seem to happen as if by "magic," deserves a similar treatment, and we will attempt just that by taking a look at the puzzle of synchronicity.

The Swiss psychiatrist Carl Jung first coined the term "synchronicity" to describe the phenomenon of seemingly unrelated events occurring in unexpected relation to each other, where coincidences are not connected by cause and effect but by simultaneity and meaning. Simply defined as "meaningful coincidence," this phenomenon is unpredictable and seemingly random in occurrence—yet it happens more often than might be supposed.

Jung described this external condition as if it paralleled the internal state of mind Csikszentmihalyi mapped. Then Jung stated that synchronicity can be experienced in one's life when, during any given period of time, two or more meaningfully related events occur without a mutually connected cause.

Interestingly, those who often remember their dreams engage in devoted prayer, meditate, or have undergone a transformation of consciousness experience synchronicity as a daily event, even routine. The phenomenon stops being a phenomenon for these people and loses any sense of randomness. It becomes instead a component of daily activities, signaling the emergence of more rhythmic and harmonious life patterns.

Remember when my car was hit on September 11 at 11:00 p.m. and cost $111 to fix? This is an example of Jung's definition of synchronicity as a meaningful coincidence. The repeating elevens eased my anxiety at the time, convincing me that whatever or whoever was in charge of my trip was still in charge and that the four-day layover that was supposed to happen would indeed happen.

Jean Munzer, a friend of mine, decided to give meditation classes in her home and had just discussed how kundalini works (that "mechanism" for releasing spiritual power, said to reside at the base of a person's spine in the shape of a coiled serpent, which stretches full length up the spine as the individual becomes more enlightened). As her class began to meditate, a real snake slithered across the lawn and stretched full length up her glass patio door and peered in, as if on cue. As you might imagine, her students were awestruck. The added emphasis of the snake's timing made the class unforgettable.

There is no cause-and-effect relationship between events in either of these two cases, yet meaningfulness is obvious.

Such is the pattern of synchronicity—little things, ordinary things, everyday things, so magically connected together that life is enlivened by the very meeting that joins them, fostering consequences of heightened importance. An understanding is reached or a message is recognized or a question is answered or a lesson is learned or a thought is expressed or a truth is revealed. The impossible proves possible, and it happened to you but you didn't cause it. Nor did anyone or anything else, at least not that you can discern.

Synchronicity catches your attention and gives you a lift and unveils, by the very fact that it exists, an underlying orderliness to the universe.

Take several nights ago when Terry was busy with studies on corporate mediation techniques so he could qualify for certification as a mediator. For no special reason, he suddenly reached over and flipped on the radio. Much to his surprise an hour-long program on mediation techniques was just beginning—a program that covered exactly what he needed to know at the exact moment he needed to know it.

And the time Terry was standing in the checkout line at a grocery store when he casually picked up a copy of the *National Examiner* tabloid. The paper fell open to a full-page spread entitled "Millions of Americans Can Foretell the Future"—featuring me and the research I was doing on future memory. Imagine his surprise in discovering that article; then imagine my surprise. We were never able to trace exactly how the tabloid got the story, nor would we have known such an article was ever published had it not been for Terry desiring something to read as he stood in line waiting for a cashier.

Synchronicity?

Perhaps.

Recall again what happened to me in 1978 when I decided to leave Idaho and trek thirty-three days across five thousand miles and through thirteen states. When meaningful coincidences are almost constant twenty-four hours a day, as they were for me then, are they still considered "meaningful coincidences"? Can we even use the term "synchronicity" to describe an almost endless string of them? Or, is there yet another aspect to this phenomenon that has been overlooked?

How my book *Coming Back to Life* evolved is a case in point. Since the entire story is too long to convey, here are a few highlights. Notice the repetitive threes and reoccurring themes of death/birth/rebirth:

- My friend Wabun's mother died, so as a way to help deal with her grief, she asked me to write three articles about my three near-death experiences—articles she published in *Many Smokes* magazine (known today as *Wildfire*).[16]

- Because so many reprints were requested, I later combined the three articles into a small book entitled *I Died Three Times in 1977*, which was purchased in a Connecticut bookstore by Kenneth Ring, a leading researcher of the near-death experience. He phoned, and at his urging, I began to write articles on a regular basis for *Vital Signs* magazine, detailing what I had noticed about the aftereffects. These articles were the foundation of *Coming Back to Life*.

- Three people at Dodd, Mead and Company (the hardback publisher) lost a loved one while working on the book and took solace in what the book had to say. Just before a third printing could be ordered, the company "died," and my royalties died with it, because of a corporate takeover—on the very day payments on two loans Terry and I had taken out to help promote the book came due. A third loan was necessary so I could buy back my world rights plus the three subsidiary contracts the company had negotiated. This situation nearly bankrupted us. The book was "resurrected" the following year in paperback by Ballantine Books. Literally, the book about the near-death experience *had a near-death experience*!

- Three months after I first started promoting the book, I was in Twin Falls, Idaho, the city of my birth, as a guest of L. James Koutnik on his radio show. After the show, he confided to me that he was ready to die. Three months later he discovered I would soon be on the nationally televised *Geraldo* show; he announced this to his audience and died three hours later. L. James helped with my "birth" into the world of business after I started my first job. When he needed the favor returned, my book was there to help him prepare for death without fear.

- During meditation one day, I "saw" what the book's cover would look like when published and "heard" the words, "Let Dick Canby do the cover." It took three months and three people before I could find out who he was (a photographer who did book covers), and contact him. Three days after receiving my brief note, he replied by saying: "I find it very interesting you received my name in meditation to do the cover for a book about the near-death experience because I, too, am a near-death survivor." His photograph on the hardback edition was the same one that I had previously "seen," even though I never saw any of his submissions, nor did I have any part in the selection process.

- In many ways, my research on near-death validates the work of Richard Maurice Bucke, M.D., a Canadian psychiatrist, and his classic book about the aftereffects of spiritual enlightenment called *Cosmic Consciousness*.[17] He and I were born one hundred years apart and we each, because of a sudden transformational event, sought out others in an attempt to better understand what had happened to us. We were later inspired to utilize our personal quest as a starting point for further research. We both discovered a universal pattern to the aftereffects we noticed, offering tangible evidence to support the validity of the transformation process. And what we discovered was virtually the same. Bucke was puzzled as to why this sort of thing never seemed to happen to women, yet it was me, a woman, who replicated his work eighty-eight years later in the year 1988.

There's more.

My three brushes with death that resulted in three near-death experiences were initially caused by a miscarriage. The man who impregnated me threatened suicide after my near-suicide when I "died" the third time (explained in detail in chapter two of *Coming Back to Life*). Three years into the book's writing, I developed writer's block, so Terry and I took time off for my son Kelly's wedding in Boise, Idaho. Our camera malfunctioned (impressing upon me that the marriage was not to be) at the same moment my menses began a flow of black blood, which continued for three months. When the doctor ordered a hysterectomy, I told everyone I was going to "birth my uterus." Because of overcrowding I was assigned to the maternity ward (a perfect place to give "birth") and given the room donated to the hospital by the radio station where my husband Terry worked (Admissions knew nothing about the room's history). The third day after surgery, I was walking past the newborns when I glanced at a large wall calendar which read "May 2" in black letters and a clock with black hands that pointed to 2:22 p.m.—*the exact day and time* I had birthed my first child, Kelly, nearly three decades before. Three months later, while driving to Blacksburg to deliver a talk on death and dying, the blocked book inside me burst forth in a gush of radiant light, complete with chapter headings and direction on how to write it. After I returned home, the call came through about Kelly. His wife of three months had walked out, sued for annulment, and obtained an abortion. He was distraught.

Kelly's energy both opened and closed my womb. I brought him into this world and he returned me to it, for it was he who found my body after death number three and "called" me back with his voice. My surgery was precipitated by his wedding, a surgery that freed me from the flow of blood to the flow of radiance—which corrected the book. Three years later I was at his side when he bottomed out from the black abyss of despair that nearly drove him to suicide following the annulment.

Three death events, three near-death episodes, three near-suicides from each of the three people involved in my third near-death, and three pregnancies—for on the day I told my youngest daughter, Paulie, I had finished the book, she informed me she was pregnant.

That first grandchild, born at 10:31 p.m. on October 31 (10/31), was pictured with his mother and me (three of us) on the back cover of the hardback edition in a photo taken three days after the baby's birth.

Seldom did I ever experience synchronicity before 1977. Since then it has become almost constant. Does this mean my life has changed? Or, does this mean my perception of life has changed?

Clearly, there's more to flow than either Csikszentmihalyi or Jung described.

Another Look at the Various Aspects of Flow

Perhaps some basic science can help us reconsider this whole issue. With light waves, it takes only one-tenth of 1 percent in any given grouping to become a dominant force; then all other light waves in that group will follow suit, creating a pattern of coherence (two or more sticking together in unison).

You can demonstrate this dominance formula for creating patterns of coherence by swinging one suspended steel beam in a room full of them. It doesn't take long before all other steel beams are swinging in unison with the first—behaving exactly like the light waves. Women living together offer another example of this formula for, eventually, all will experience their menses during the same cycle as the one whose presence dominates. Public moods and trends follow the same pattern. So do mobs.

This natural tendency to cohere (hold together) in unison with whatever urge or force is dominant at the time is called the law of resonance. It was discovered quite by accident in 1665 when a Dutch scientist observed that in a room full of clocks with pendulums swinging at different tempos in different directions, soon all pendulums would swing at the same tempo in the same direction as the first to swing.

Thus, the law of resonance describes patterns of relationships in motion. It is where disconnected parts connect to form a whole based solely on how each part is able to respond to and relate to the others (resonate).

According to science, this pattern of relationships in motion can be formed two ways: through manipulation (where it is purposefully caused), or because of the influence of a dominant member or part (as with mobs, trends, pendulums, women menstruating).

Yet there is a third way—*through the connectedness one feels when consciousness is released from the constraints of thinking.*

When consciousness is allowed to function by itself, without any demands on its thinking aspects, it freely floats into a flow state. My experience has been that, regardless of which condition of flow results, that particular state will automatically attract to it whatever else matches its harmonic resonance. This enables the different parts of that vibratory pattern (whatever fits) to cohere (come together) in relation to each other (that which fits together comes together). And this produces patterns of relationships in motion, *in accordance with the law of resonance*.

In other words, the law of resonance applies the same, whether addressing an internal state of consciousness or external events that happen in the physical environment. And the dominant force that brings about coherence during any aspect of flow (inner or outer) is a power and a presence beyond the personality self of any individual—a greater stream of intelligence.

What's been said thus far can be charted:

ASPECTS OF FLOW

INTERNAL TO SELF	EXTERNAL TO SELF
Subjective environment	Objective environment
Without a focus	More in focus
Release of thoughts	Release of goals or vested interests
Stimuli fades away	Stimuli increases in clarity
Blank out to nothing	Perk up to new possibilities
Consciousness expands	Experience expands
The mind flows	The life flows
You know more	You do more
Connect with a source of wisdom greater than self	Connect with a source of guidance beyond self
Gain information	Gain harmony and an orderly rhythm to life experiences
Unify in consciousness	Unify with the world at large
A state of mind	A state of being

The only purpose of synchronicity is to catch our attention. It signals that an external flow has been activated (for however long).

The Puzzle of Synchronicity Revealed

Simply put, synchronicity is the visible signal that shows us we have connected to and aligned within a greater stream of intelligence, a wholeness beyond ourselves—perhaps even the true source of our inner being. When we are "on course," even for a moment, synchronicity tells us so. It gives us the feedback we need to recognize that an alignment has occurred.

I interpret the activation formula of one-tenth of 1 percent to indicate that it doesn't take much effort on our part to initiate the resonance of relationships in motion (internally or externally). When we are receptive (relaxed, willing, open to receive), the mind is released to seek out more of our potential and access a larger vision. (This addresses both positive and negative aspects, as it illustrates how a single individual can sway the masses.)

I know of no one who operates functionally in the earthplane who can maintain continuous internal flow states. We need the force of our own will to motivate the "unfoldment" of our lives and guarantee that we experience and learn while embodied. Flow states enhance life; they do not offer a way to escape it. I do know of many people, however, who experience external conditions of flow quite often, some almost continuously, as if such a pattern of relationships resonating together while in motion was a natural by-product of their commitment to a more balanced and spiritual lifestyle.

Our societal notion of "coincidence," at least to my way of thinking, is actually a joke we play on ourselves. We use the term as a catchall so we can avoid explaining what we do not know how to explain. I have discovered that there are no coincidences in life, just gaps in our ability to understand how unrelated parts can connect to form a unified and meaningful whole.

Future memory is similar in how it occurs to the varied aspects of flow, except it is more pronounced and intensely felt—and *it can be willfully manipulated.*

9

Memory Mazes and the Brain

Time and space are modes by which we
think and not conditions in which we live.
—Albert Einstein

At this point we know that future memory is the ability to prelive future time while still active and functional in present time. The experiencer feels as though he or she has slipped through a crack in consciousness when this happens, a crack that links one into a flow state beyond the self (a greater stream of intelligence). Preliving the future involves sensory response and physical activity so detailed as to emulate the normal life routines of present-tense reality. You are aware that something different is occurring only because of the exhilaration or emotional thrill that accompanies the onset of the experience. The event usually passes into realms of memory soon after, since the demands of daily living take precedent. Later on, though, the memory resurfaces. A certain scene, sound, feeling, or smell triggers it. Once the memory of a future already lived returns, there is no doubt of its validity. You know that you have previously lived what you are just now beginning to experience. This knowing may seem funny, strange, or puzzling, yet the brain registers it as fact, not fantasy.

You remember the future because the brain, your brain, can register and process any type of memory with equal efficiency. And that's because the brain routinely notices and combines and creates and remembers images and activities throughout all its varied awareness modes.

Let me reword what I have just said: Fiction and nonfiction can both be true in the sense that subjective reality and objective reality are processed the same in our brain.

Yet in our society, the conscious state of brain function is valued most, with the subconscious hardly taken seriously. This is unfortunate, since while the conscious part of our mental capacity separates and dissects and measures to arrive at single units, it is the subconscious aspect that extends, enhances, and reconnects to form more meaningful perspectives of the whole. But when we enter into a flow state, we access the subconscious directly. It then becomes our "launching pad" to destinations deeper within or beyond what the brain typically perceives.

Both conscious and subconscious states of awareness are commonly associated with specific brain hemispheres; these hemispheres can be trained to learn language and engage in thought processing. The language of the right brain hemisphere (related to the subconscious) is symbolic imagery, and it follows the abstract of intuitive knowing. The language of the left hemisphere (conscious state) is detailed glyphs, like letters and numbers, and it follows sequential ordering and deductive reasoning. Each brain hemisphere performs functions on par with the other.

There is a third major area in the brain/mind assembly, however, and it includes the temporal lobes and the neocortex (the higher brain), both directed by the limbic system. The limbic is a conglomerate of various parts and sections located in a semicircle in the middle of the brain and capping off the topmost extension of the brainstem. It wraps around our primitive reptilian brain, translating our basic instincts for sex, hunger, sleep, fear, and survival into more flexible and social forms of behavior. Often referred to as our emotional or feeling center ("gut" brain), the limbic is also the seat of our immune system and the body's ability to heal itself. (Smell is the only faculty that accesses the limbic instantaneously. All other stimuli take longer to register, hence the importance of odors.)

Few people realize, though, that it is the limbic that operates as "the executive office" in deciding what information is stored in memory, what is forgotten, and what will be further elaborated upon and refined in the two main hemispheres and throughout the brain/mind assembly. And, it has a direct neural connection with the heart.

Although this small, but extremely efficient, system has been around for hundreds of thousands of years of brain evolution, only recently has the limbic been recognized as *the most complicated structure on earth*. Many professionals now believe that if the limbic does not originate "mind," it certainly is the gateway within the brain to higher realms of mind and more powerfully diverse and collective types of consciousness.

Thus, *the staging arena* where the organ called brain accesses and filters what is referred to as "mind" *is the limbic system*. Remember this.

Direct or conscious entry into the limbic section of the brain is gained through any type of excitement or heightened stimulation (outwardly expressed emotion). Indirect or subconscious entry is most commonly reached through the blissful openness, even ecstasy, experienced in altered or flow or meditative state (inwardly experienced emotion). One way or another, some form of emotion is necessary to accelerate and prolong limbic involvement (which can also activate frontal lobes and neocortex).

(Parapsychology, that branch of psychology that deals with the investigation of psychic phenomena, has discovered the same link. Example: You cannot conduct accurate and repeatable psychic experiments without some form of excitement or emotion felt and expressed by the subject. No emotion results in little or no phenomena. When there is plenty of excitement, there are plentiful results. Future memory, as you will recall, is accompanied by a "thrill" that excites.)

One more thing. The importance of the limbic is the reason why today's scientists consider the brain more emotional than cognitive. Nicholas Humphrey, a senior research fellow at Cambridge, states: "A person can be conscious without thinking anything. But a person simply cannot be conscious without feeling."[18] Add to his statement this additional scientific finding: Feedback between the limbic system and the heart *is immediate*!

We can condense and simplify correlations and patterning normal to the brain/mind assembly in this manner:

Correlations within the Brain/Mind Assembly

Left Brain Hemisphere: Consists of the conscious, objective aspect of awareness. It mainly analyzes, clarifies, categorizes, and separates. Intellect and reason are its regions of expertise; science and education its preference.

Right Brain Hemisphere: Consists of the subconscious, subjective aspect of awareness. It mainly collects, absorbs, enhances, abstracts, and connects. Imagination and intuition are its regions of expertise; religion and fantasy its preference.

The Limbic System: The gateway and guide to super-conscious, synergistic aspects of awareness (utilizing the neocortex). It mainly senses, embraces, and knows that it knows (gut response). Relationships, memory, and the collective whole are its regions of expertise; mystical knowing (gnosis) and convergence within realms beyond self (unification with a greater stream of intelligence) its preference.

This list enables us to take a brief glance at some areas of the brain/mind assembly. Not that the mind is necessarily divided into parts or that these sections of the brain do only what has been ascribed to them; but, rather, by recognizing the major states of awareness our mind has in context with brain support structures, we can better understand and appreciate the inner workings of our own being and how our awareness of what seems real naturally alters as consciousness shifts its mode of operation.

And these brain support structures are utterly amazing. For instance, science estimates that the human brain can store in the neighborhood of one hundred trillion bits of information. A computer by comparison, even the best, can only store a few billion. *And the brain can rearrange itself in fifteen minutes or less.* Experiments with PET scanners (the machine that shows in color which regions of the brain are activated when used) have shown that original thinking "lights up" areas of the brain not used for any other mental activity. This proves that the more novel the task, the more significantly brain function and brain structure alter. Thus, time is not a factor when shifts occur in the brain.

Actually, the only limit to the brain's ability to process and learn new information is the willingness of an individual to use his or her mind creatively.

Yet, since the brain organ itself cannot tell the difference between a dream and a physical event, as both register the same in memory (remember what I said about fiction and nonfiction), *whatever is perceived during any state of consciousness is remembered.* By recalling that memory, a former state of consciousness can be *reexperienced*—past, present, *or* future.

The brain remembers.

The mind never forgets.

"But the whole body remembers, not just the brain," cautioned Peter Derks, psychology professor at the College of William and Mary and a specialist in memory research, during an interview I had with him several years ago. He put my knowledge of how memory works into perspective by commenting that, "any change in the system is a kind of memory." He explained: "A scratch or a broken bone is a 'memory' of an injury. Chronic pain, that muscular tension associated with neural (nerve) damage, is a conditioned 'memory' of the damage after it has healed. More specifically, an event or change in the environment is registered by a change in the nervous system. A pattern of environmental stimuli produces a pattern of neural responses."

Derks helped me to recognize that it is the patterning, those distinctive clusters of associated responses, that builds memory and enlivens memory recall. And, as current research has made plain, it is our emotions that strengthen our memories. (Future memory episodes are no exception, because of the exhilaration or emotional thrill that accompanies them.)

Emotions enable memory clusters (patterns of response) to hold together easier and better and longer, and be accessed quicker. And emotions are a product of the limbic system (as well as being the best way to stimulate direct access to the limbic system). And it is the limbic system that decides what is remembered and where that memory is stored, and that doesn't necessarily mean in the brain. As Derks said, memory clusters can be stored anywhere.

Whenever you refer to memory, and in whatever manner you make that reference, you are actually referring to the limbic—at least initially—*and the limbic is the jumping-off point to collective and synergistic reservoirs of knowledge and wisdom beyond the self.*

Compare again what we have been discussing with what commonly happens during future memory episodes. That emotional rush of energy, coupled with an expansion of consciousness, propels one into a realm of limitless possibility, where the future can be prelived. Experiencers report feeling as if they have slipped through a crack in consciousness, as if a gap exists between the activity of thought and the silent nothingness of no-thought—a gap that could well be located in or through the limbic gateway.

Pause for a moment and let this sink in.

While you are relaxing, be reminded that the true art of memory is the art of *attention* (directing the mind) and *intention* (exerting the mind).

The more alert we are, the deeper we can concentrate and the more information we can access, and that's what all of this is really about anyway—*information*—knowing more.

Illustration: When listening to music, don't just hear it, concentrate (the art of attention). Bring the background sounds into the foreground with your mind (the art of intention), and see how much louder the music will be and how many more nuances of sound you can distinguish. Turning up the volume puts your eardrums at risk. You need only increase your attention and intention to increase the pleasure various sounds offer and the experiences you can have (which gives you more information—you know more because you accomplished or absorbed more).

There are other ways, though, to regard memory:

- The ancient Greeks considered all learning *remembering*; to them, life was but the act of recollecting knowledge the soul forgot at the moment of birth into a physical body. Reincarnation and the idea of past lives were integral to their worldview. Perhaps this explains why the word "education" originally meant "to draw from that which was already known."

- The French coined the term "déjà vu" to describe the surfacing of past life memories (as separate from memories of the present life), or the "illusion" of having previously experienced something actually being encountered for the first time. Also known as the phenomenon of the "already seen," déjà vu can signal a rerun of projections from the dream life as well. (Déjà vu is distinctly different from future memory in that it centers around "past" and varied extensions or distortions of memory recall rather than the all-encompassing and sensory-rich experience of preliving the future.) Some medical researchers postulate that déjà vu is but a condition of skeletal memory images, incomplete, and formed only with enough detail to allow recognition and distinction. Others remind us how holographic the brain tends to be, and how easy it would be for different holographic memory clusters (with shared elements) to fool us into thinking we had done something before when we had not. The purpose of this might be to alleviate stress, like, "I've done this before and it was okay; therefore I can do it again."

And there are countless types of memory, mazes of them: in your brain, in your body; vivid memories, partial memories, wispy memories; layers of memory from this life, past lives, future lives, from varied states of consciousness, and even from beyond the self. Alertness unlocks the patterns of information stored within each type.

From mindscapes deep inside us can emerge still other kinds of memories, "rememberings of rememberings." Of this, I have a story to tell.

That November night in 1978 when I drove to Middletown, Virginia, from Washington, D.C., to deliver my initial lecture on the near-death phenomenon, I noticed something strange about how the Shenandoah River affected me. An experience akin to sticking one's finger in an electrical socket sent shock waves of energy through me as I neared the river. Although I had heard of the fabled Shenandoah before, this was the first time I had been in its environs and the physical jolt I felt from its energy baffled me. After the lecture ended, I returned to Washington, D.C., that same night, crossing the river again. Here's what I later wrote in a note to myself about that second encounter with the Shenandoah River:

> The Shenandoah made itself known well in advance of my car's headlights. But this time the energy waves felt more like ripples of symphonic music than electrical shocks. Melody crested with each wave as the sensation it gave me ebbed and flowed. My car slowed to a crawl as the bridge came into view; then, most unexpectedly, I floated free of my body.
>
> Leaving my car far behind, I rose high into the air, hovering, then plunged into the water's cradle, merging with rocks and pebbles and particles of soil, fish, and water molecules. Ever so slowly my expanse grew larger until the largeness of me encompassed riverbanks, weeds, grasses, bushes, trees. Every pulse anything made registered as my pulse; every touch and sensation spread throughout my being in an orgasm of unbelievable pleasure. Up the trees I rose to caress again the air and, in free flight, I watched each droplet and particle flicker as an atom in my own body. The land, the waters, the air, the plants, and all that dwelt therein unified within me and I beheld myself as the very substance that held together this fragile portion of earth. I was its guardian, its Spirit Keeper.[19]
>
> This staggeringly vivid awareness of guardianship unveiled itself as the presence of my own soul, that greater part of me still directly connected to God, still pure and untarnished and without personality or ego. My soul was the fullness of truth, one of countless reflections God uses to recognize Itself looking back at Itself. And as a Spirit Keeper entrusted with this particular speck of creation, I also beheld other souls, uncounted numbers, each holding together that part of creation entrusted to them. This was our job, our privilege, to keep matter bonded within the spirit which gave it form.

I saw that even while embodied as humans we souls remain partially detached, free to exist beyond what seems our limits, free to serve the ongoing flow of cosmic breath in whatever capacity needed. We are existent in a body while equally existent elsewhere, simultaneously, because what we refer to as existence, life, is more dynamic and flexible than any form can encapsulate. By coming to the Shenandoah River, I had returned to where the greater portion of me already existed. The shock waves I first felt were caused when my persona of Phyllis merged within my own soul mass. The sensation of this reunion was bliss beyond bliss, ecstasy beyond ecstasy.

How long I remained in this state of utter joy I do not know, but when I again became aware of my car and of me sitting behind the wheel, I had long since driven past the Shenandoah River and was well on my way back to Washington, D.C. Then, in hardly more than a wink, I found myself walking in the door of my apartment so transformed and so refreshed that sleep was unnecessary.

Throughout what hours remained until morning, I pondered these questions: What if déjà vu is actually the experience of an embodied soul encountering the presence of the rest of itself? What if déjà vu has more to do with the reunion of the lesser self with the greater self (our soul mass) than anything associated with memories of past lives? What if déjà vu accurately describes the sensation of reconnecting with where part of our soul mass already exists, where we as Spirit Keepers are guardians of the creation we help to hold in place? What if Spirit Keepers are that portion of individual souls comprising the very substance manifestation adheres to and shapes itself around? What if creation exists because we the created needed a place where we could become aware of ourselves as God experiencing God through us as us?

I greeted the new day with the conviction that it was time for me to seek out other near-death survivors and ask a lot of questions.

I want to emphasize that what happened to me at the Shenandoah River that night was not in any way a typical out-of-body experience. True, I had certainly left my body far behind, but, as a former teacher of out-of-body states and astral (otherworldly) traveling back in the sixties and early seventies, I can say that this experience was most assuredly different. It had the intense power and the textured richness of a near-death experience, plus the revelatory awareness of encountering a grander form of identity. But more importantly, there was an instantaneous resonance with a type of consciousness and an energetic form of remembering previously unknown to me.

As my lesser self merged with this aspect of The Greater Self, my awareness shifted to one of remembering my own remembering. Clearly, it was memory at work, not discovery. Although I was assailed with serious questions after the experience was over and I had returned to Washington, D.C., during the actual reunion of selves, this "remembering" connected me with fields upon fields of layers upon layers of collective or mass memories. It was as if by joining with another aspect of "the self I really was," and remembering Its memories, I automatically resonated with levels of memory existent beyond anything I had ever known—whole "libraries" of them, wave after wave of them—as if everything that existed had its own field array of information storage about its own existence, and those fields ("memory libraries") were accessible when one resonated with them.

It was a dazzling experience, one that changed how I regard memory, what it is, and how it functions. It also changed my awareness of how the mind can work and the interplay between objectivity and subjectivity and realms beyond the kind of "self" we see when we look in a mirror.

If you take the section entitled "Correlations within the Brain/Mind Assembly," found earlier in this chapter, and reread it as levels of awareness (instead of brain/mind correlations), this pattern emerges:

Conscious: Normal ego awareness
Subconscious: Altered states of awareness
Superconscious: Enlargements of awareness

Do the same thing again, only as states of existence:

Objective/Physical: Wide awake, alert, externalized; outer world
Subjective/Symbolic: Dreamy, subliminal, internalized; inner world
Convergence/Synergy: Mystical, knowing, unified; collective whole

How our mind responds is a reflection of how easily our mind operates through the various brain support structures. What state of existence we experience at any given moment depends on the mode of awareness our mind is in at that time.

Of continued intrigue, however, is the puzzle of what else might the mind be capable of and what range it might need to cover for a phenomenon like future memory to exist.

Of all the present-day thinkers yet to explore this puzzle, I think Itzhak Bentov was the most innovative. His books are classics. Using humor as a tool, this biomedical engineer turned cosmologist tripped across the limitless boundaries of mind and memory and made some startling discoveries outlined in his ever popular *Stalking the Wild Pendulum: On the Mechanics of Consciousness.*[20] After his untimely death from a plane crash in 1979, his widow, Mirtala, pieced together enough material gleaned from talks he had given along with notes left behind to produce a sequel on the mechanics of creation, entitled simply *A Cosmic Book.* Bentov died before his cosmology could be completed, but it is obvious to me that what he did develop presents a workable framework for continued research on the subject.

With permission from Mirtala, here is one of Bentov's diagrams we can use that illustrates his thoughts on objective, subjective, and convergent time-space relationships:

Itzhak Bentov's Version of Objective, Subjective, and Convergent Time-Space Relationships

Normal State

Both versions of time parallel each other; both versions of space parallel each other. Time appears as sequential and orderly, while space appears as air filled with solid objects. Our environment and all that we perceive are normal, ordinary, and what we are used to. Nothing changes. The integrity of perception is unchallenged.

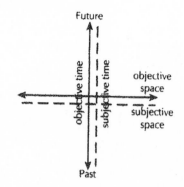

Altered State

Subjective time begins to separate from objective time. The same separation occurs with subjective and objective space. Time is no longer sequential and orderly. Space changes and objects appear less solid and dependable. The space-time coordinates we are used to lose their relevance. Our perceptual modes of awareness are freed to expand and enhance, while sensory feedback either decelerates or accelerates.

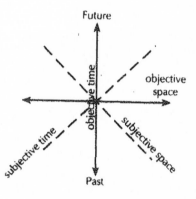

Convergence

Subjective space overlaps with objective time; objective space overlaps with subjective time. Everything converges, making it possible to be everywhere at once, all at the same time. There are no longer any separations. Nothing divides. Normal coordinates and definitions disappear altogether. Limitations of any kind no longer exist.

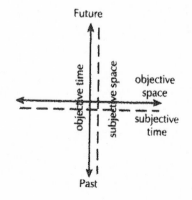

During an interview I had with her, Mirtala explained that Bentov felt there were two kinds of time and space: objective (conscious), existing in verifiable units that could be scientifically measured; and subjective (subconscious), more flexible and capable of being stretched or compressed according to one's own state of mind. He is quoted as saying:

> We know that in an altered state of consciousness, such as a dream, we can experience an enormous amount of events in only a few minutes. Time seems to expand, and for every objective second one has hundreds of subjective seconds.
>
> Now, imagine a situation in which the subjective time axis deviates so far that it becomes parallel to and overlaps objective space. This would imply that our subjective time fills all objective space—it takes no time to go anywhere. This is what happens in expansion of consciousness: We can fill the entire universe with our consciousness at infinite speed and be everywhere at once, in other words, become omnipresent.

The concept of omnipresence, as expressed by Bentov, is a condition whereby all forms of existence, time and space as well as objective and subjective, overlap and then converge. Consciousness is not only freed when this happens, it can expand to fill the whole of creation, instantaneously.

The result? "Everywhen."

(It took a poet like Gary Snyder to clarify what science still seeks to identify. In case you haven't noticed, the quantum theories of subatomic physics are easier for poets to grasp than for scientists.)

Many researchers have taken Bentov's ideas and developed them further, hoping to discover the range of mind and memories. One of these people is Jack Houck, a systems engineer in the aerospace industry.

Houck originated what came to be called PK Parties, "PK" being an abbreviation for psychokinesis or "mind over matter." During these parties, Houck's goal is to teach people how to bend metal using a combination of mental images and commands along with gently rubbing or "warming" the item to be bent. He claims that the secret to success lies in how relaxed yet knowingly expectant a participant can become (emotionally receptive). During a meeting I had with him in Washington,

D.C., he maintained that once the right state of mind and excitement level are achieved, an item will bend or twist with the slightest touch (sometimes with no further touch at all), be that item silverware, keys, or metal tubing. Over ten thousand people have attended these parties since 1981, with two-thirds of them bending the metal item they furnished themselves.

From this large cross-section of people and from many other research experiments he has conducted, Houck put together his findings about consciousness in numerous professional papers, including "Conceptual Model of Paranormal Phenomena," which was originally published by *Archaeus* magazine.[21]

Concerning memory, Houck explained to me that the mind functions much like a sophisticated radio that searches for a peak radio-signal intensity. Once that frequency is found, the radio will lock onto it, enabling us to hear what's there. He noted that, in this same manner, the human mind also searches for peak intensities, emotional ones, and once found, the mind, too, will lock in place, thus enabling the data to be retrieved. This data can be of video quality, quite detailed, and involve all the senses. This retrieval process is similar to holography, where one part unveils the whole.

Houck is one of many researchers who discovered the same thing I did early in my career: If you want experiments to work in the field of the paranormal, you must increase the emotional intensity associated with the current time the event is in progress. Without that emotional intensity, the mind will not scan, store, or lock onto anything in particular, especially the sensory-rich memory clusters, and nothing much will happen.

And that means you must involve the limbic system.

But this systems engineer turned brain/mind researcher had more to say about consciousness and the process of memory making during our meeting, and his additional comments are important.

According to Houck, once you access a particular memory, you can see, hear, and feel the information of that event from your past. Similarly, when you perceive possible future scenarios, you can see, hear, and feel that futuristic information much the same as when you accessed a past memory. He explained that when you look into the future, you temporarily collapse the "wave mechanics" of everything associated with that future event (in physics, the mechanics of existence and creation are described as being dual in nature—behaving as waves while appearing as particles). You then store this information about the future event in your portion of what "mind" may be (that part of "mind" accessed and filtered through your brain), with the information later becoming available as memory. (Refer to appendix II for a diagram of Houck's that illustrates this concept.)

He further stated that

As you move forward in time, there might be changes in the wave functions for that future event, because people may change their decisions about something. When you get to that perceived future event, it may be changed. An example is the case where you may look forward in time and see yourself at some intersection. Then down the street comes a yellow car. As you step off the curb, you get hit and killed. You may not like that. In the future, you may happen to arrive at that intersection and have a "déjà vu experience." This time you see the yellow car coming down that street, and, as you are stepping off the curb, you can either go ahead and get hit, or use your own will power and step back up on the curb. If you step back, all the probable futures for everyone and everything associated with that event have to be reshuffled to become new sets of wave functions.

Both Bentov and Houck describe a "simultaneous-everywhere" model of consciousness, where all information preexists and can be accessed from any point in the universe at any moment by anyone. Their work indicates that our formerly accepted notion of time-space coordinates and sequential ordering needs to be updated by a more holographic understanding (the idea of a simultaneous-everywhere grid of information, where any part can recreate the whole).

This indicates that the reason future memory can exist is because future potentiality preexists.

Repeatedly, various types of research have shown that the seat of memory formation and memory retrieval, whether simultaneous or sequential, lies within the limbic system, and *the limbic system is the information gateway to the higher mind and realms beyond the self.*

Intuition is the equal of intellect in the limbic, and the only way out is in.

10

Illusions of Perception

Each of us makes his own weather, determines the color of the skies in the emotional universe which he inhabits.
—Bishop Fulton J. Sheen

One man who experienced regular episodes of preliving the future stated that reality seemed to him as if an echo from some primal movement in time and space, and that the opportunity to live advancements of time was an indication that those echos were of differing wavelengths and sizes. "When we slip through the waves," he conjectured, "we are able to experience reality from a different vantage point, literally from another frequency of vibration. It is the focus of our awareness, our perception, that determines what frequency we pick up. It all seems real because it is real."

Personal testimony like this, along with the exciting research currently being conducted, gives us pause. Yet the fact is, we must make the judgment call for ourselves when the status of real versus unreal is subject to question. The existence of what we can see, hear, touch, taste, feel, and smell is undeniable—or so we think. Still, the phenomenon·of future memory refutes this. As the parable of the five blind men and the elephant clarifies, what appears to be absolutely certain isn't necessarily as certain as it seems. Our faculty of perception can fool us.

What then determines real or unreal?

Lawrence LeShan, in his classic work *Alternate Realities: The Search for the Full Human Being* says: "A reality is real to you when you act in terms of it. Anything else is just talk. It is a valid reality when, using it, you can accomplish the goals acceptable to it. Common sense rules every reality and ultimately decides on its validity."[22]

LeShan's statement reflects a discovery made by a team of scientists who were experimenting with babies. They found that the only time babies startled was when something happened to them that defied common sense. This discovery established that a certain level of perceptual prejudice is part of our genetic predisposition, a predisposition reinforced by our various faculties and our brain. We depend on life being what we think it is, and we accept the bias of that perception. Throughout day-to-day existence, *we recognize only what we are prepared in advance to see.*

Alternate realities and other dimensions of vibration are missed or bypassed simply because we are not aware that we are missing or bypassing anything. We accept what we perceive, and it seems illogical, if not impossible, to do otherwise.

But this tightly knit package of natural perceptual prejudice (often referred to as "environmental integrity") is actually based more on assumptions from individual belief systems than on genetic predisposition. Often, it is more a *preference* than a prejudice.

This is so because of the way we mix together acquired tendencies with natural perceptive skills. We allow our loved ones, our schools, our jobs, our fellows, our society, our governments, not to mention our own perceptions of what we think we perceive, to define and interpret our lives. We allow this because it is fundamentally easier, more practical, and less risky, to accept rather than deny the bias of mutually accepted belief. (Society owes its existence to this tendency among people to accept majority opinion as personal truth. Messiahs owe their deaths to the same principle.)

To get at the heart of this issue, three examples of natural perceptual prejudice follow. A fourth is presented at the close of this chapter. Pay close attention to the paradoxical illusions each example unveils:

Example 1: You go to a movie (formerly known as "motion pictures") to enjoy a good show, but what is it you really see? Quite literally the continuous projection of a series of still frames separated by periods of darkness. It is your *perception* of what you think you see that supplies what appears to you as the movement of a solid storyline. Nothing you see in itself is capable of either movement or coherence until you, the viewer, supply both by connecting what the projector projects within your own mind. What you think you see doesn't really exist. Only the continuous sequence of single units exist. It is your mind that connects them. Movies are an optical illusion.

Example 2: You sit down in front of your television set to enjoy a good program, but what is it you really watch? Quite literally one electron at a time (with black-and-white, and three at a time with color) fired from the back of the television tube to the screen to be illuminated once it hits the screen as a tiny dot. The continuous barrage of electrons-turned-into-dots creates the appearance of images, as scanning lines (raster bars) roll from top to bottom separating information coming in (new dots) from information fading out (old dots). You adjust the vertical hold on your set, not to remove strange bars appearing in the picture, but to place all screen activity within the range of your own *perceptual preference*. A television picture tube is nothing more than a "gun" that fires electrons at a screen. Your mind connects the electron dots into the picture images you think you see, while it totally ignores the true reality of what actually appears. Television is a mental illusion.

Example 3: You go to a concert to hear good music, but what is it you really hear? Quite literally a series of notes separated from each other by intervals of silence. All any instrument or voice can produce is single sounds, one at a time. It is the perception of the listener that supplies melodic sweep or dissonance, what is termed music or noise. Without the listener's participation and his or her perception of what is heard, sound would be incapable of what appears to be a flow. What we hear as continuous sound is a creation within our own mind. Music is an auditory illusion.

Because we are not prepared in advance to see *through* the illusions of perception, we accept what we perceive as the full truth of what is there. Reality, in the strictest sense, is a product of our own creation and is maintained by our own perception.

The issue of *realness* can be tricky, though.

Certainly, the subconscious mind regularly absorbs more than a billion pieces of information per second. Add to this figure the fact that the average person today perceives sixty-five thousand more bits of information and stimuli per waking day than did his or her forebears just a century ago. Indeed, our brains are now so bombarded that less than 1 percent of what comes in ever reaches the conscious mind. Within a fraction of a second, over 99 percent is filtered out.

The area within the brain/mind assembly that does the filtering is the reticular activating system, a small bundle of densely packed nerve cells located in the central core of the brain stem below the limbic system. What directs the filtering, though, is *perceptual preference*—not necessarily inborn perceptual prejudice.

Yet, neither our natural predisposition nor the preferences we acquire as we mature need to prevent us from the fullness of true perception that is possible for each of us to attain. What is automatic, even from infancy, can be altered, expanded, enhanced, or changed.

Remember the children's story about the emperor and his new clothes? The tale concerns an emperor who was tricked by con artists into buying "invisible" apparel, which was then fitted and tailored with imaginary flair. Since the emperor believed the phony story as told him, none of his subjects dared contradict for fear of what they assumed the emperor might do to them if they did. A public parade was later arranged so the emperor could show off his new "finery." As he strutted amongst the crowd, one youngster recognized the truth of the situation and shouted: "Hey, look, the emperor's not wearing any clothes!" (Children, by the way, have the least amount of learned perceptual preferences blocking true perception; hence they have the clearest minds. They confront situations directly, not indirectly.)

Slowly, throughout our lives, we accept, decide, and viscerally integrate structural thought models of what we will believe and what we will reject. These thought models (perceptual preferences) create the filters (densely packed nerve cells) that prevent us from becoming aware of what we do not want to know. Like the emperor and his subjects, each accepted a particular reality as true and rejected any other alternative.

Genetically speaking, this filtering of input operates like a shutoff valve in how it gives the conscious mind an opportunity to play "catch up," so it can sift and sort through a hodgepodge of information while assessing value and worth. Without such filtering, we would surely be inefficient and ineffective; we could neither decipher nor decide, nor could we focus our attention.

And that's the catch.

We can overdo it. We can block out more than we need to. We can create so many blind spots we become as if blind, or in a trance or half asleep or locked into various stereotypes of foolishness and bigotry. We can deceive ourselves. Daniel Goleman, Ph.D., tackled this situation in *Vital Lies, Simple Truths*, where he notes: "The great antidote for delusion is insight, which is simply seeing things as they are."[23] Like the youngster in the crowd yelling at the emperor and speaking the truth others chose to ignore, we benefit when a fresh viewpoint is offered and a new challenge is met.

We need our natural genetic predisposition to perceive the continuity of motion and the cohesion of form, so relationships and comparisons can be made. We even need the bias of mutually accepted beliefs, because these very preferences and prejudices provide the filters that allow enough time and space for us to develop social skills. But we don't want too many or too much.

This means we would be wise, each one of us, to inventory our filters (accepted beliefs) periodically, reevaluate them, and consciously decide whether or not each is still operating in our best interest. We may find by doing this that some are not only outmoded and outdated, but were never really needed to begin with. As Ralph Waldo Emerson, the famous poet and philosopher, once said, "A foolish consistency is the hobgoblin of little minds."

But what about the solid realness of ordinary reality?

Yes, the examples given thus far illustrate that our faculties enforce the appearance of a solid environment. Yes, we can retrain our perceptual skills, widen them, so more territory can be included in our worldview. Yes, we can reassess and then release outworn and outdated preferences and belief systems.

But when you kick a chair, your toe still hurts. What appears as solid feels solid, and responds accordingly.

Still, the nagging question remains: is solid *really* solid?

Meditation and other practices similar to it help us to retrain our perception so the sequences of both motion and rest (described in each of the three examples given earlier in this chapter) can be viewed *simultaneously and separately at the same time*. A near-death experience or a spiritual awakening shifts the capacity of our brain even further. Such a brain shift "cleans out" our filters, blocks, and beliefs in such a way as to enable us to "slip between the cracks" of perception into alternate realities, parallel realities, and coexistent realities, the likes of science fiction. This convergence of information (chaos) is disorienting at first, but eventually we are led to that wellspring of clarity and insight formerly masked by our inborn predispositions and our acquired perceptual filters.

Let me illustrate what I'm saying: A few months after my near-death episodes occurred in 1977 I began to experience sensory input unlike anything I was accustomed to (including the synthesia, or multiple sensing, I had throughout my youth). At that time, phlebitis and the damage done by blood clots and other physical traumas required that I relearn how to crawl, stand, walk, climb stairs, as well as run. Therapeutic exercises were ongoing. A letter I wrote then describes a particular sunny day in downtown Boise, Idaho, when I could at last run an entire city block without falling and without pain. Note the sensory alterations that accompanied this feat:

> Each minute sensation from my legs was received in my brain as if it were the afterclap from a sonic boom. That loud, and I could both hear and feel simultaneously. If I couldn't hear a sensation then I couldn't feel it either because, for some reason unbeknownst to me, both faculties

had merged. They were now equal halves of the same sensory mechanism, reverberating in shouts of feeling/sound throughout my body.

As I cried out for the joy of being able to run again, I noticed rays of energy protruding from me and spiraling out into the air. They looked like pulsating flares glinting in the sunlight. A car honked when I wobbled off the curb into the street, feeling somewhat dazed and giddy. I jumped back and when I did, those energy flares flipped into fireworks, setting off a cascade of what appeared to be miniature rockets shooting off in all directions.

I could taste it, the sun, and I could taste the satisfaction of being there standing on the sidewalk. Whatever I saw or thought about deeply had flavor, a taste. My faculties for sight, thought, and taste had also merged. Feeling/sound. Flavored sight and thought. Who in their right mind would believe any of this? Me? Anyone?

My tears of joy at being able to run rolled into wracking sobs that day, for I was overwhelmed by the strange sensing multiples that assaulted my brain. This wasn't the first time since my near-death episodes that the sensory stimuli I received did not match either perceptual conditioning I was used to or what I had experienced throughout my youth. Still, this incident was a turning point for me, because it forced me to realize that more than my body needed retraining.

I have come to believe that the extremes in sensory distortions I had to deal with during this initial period after dying thrice over were the result of losing much of my inborn perceptual prejudice. I now recognize that the strange sounds I heard and the energy flares I saw were, in all probability, a magnification of biological processes normally not discernible to conscious awareness, mine or anyone else's. This magnification made my world seem oddly different when, I suspect, it was really my perception of my world that had shifted the most. It could well be that my reticular activating system might have been damaged; certainly my limbic functioning was stimulated or perhaps altered in some manner.

Regardless of cause, these novelties of perception eventually worked to my advantage in how they enabled me to enhance awarenesses beyond what was normal for me. After I learned how to control them (along with the other sensory multiples that emerged), and apply common sense in their use, my life was enriched immeasurably.

Once your consciousness transforms, whether sensing processes magnify, as I believe they did for me, or whatever else begins to shift around, the very first thing you lose is a sense of time and the second is a sense of space. As I said in chapter 2, the world reorders itself, and you find that you are no longer as influenced by the paradox of perceptual illusions.

You come to realize that solid is not really solid.

This is a dramatic switch in perception, and one I want to discuss further. Using science as an aid, here are a few illustrations of what might be taking place when time and space become illusory to one's perception.

In Newtonian physics, it is known that all manifestations of energy create time by their vibration and space by their wavelengths, that time and space are properties of energy. Where there is no energy, there is no time and there is no space. There is "no-thing."

Here's what is offered in science as a classic explanation for this phenomenon: the repetitious cycles by which energy vibrates are what create what we call time. When energy vibrates in a continuous fashion, forces within it separate as two opposing poles of attraction. The attraction between these poles causes energy to move back and forth from one pole to the other in an oscillating movement. This oscillation creates a sine wave (like a curved line or arc—considered in physics to be the most basic of all wave forms), and the length of that sine wave between the poles is what we call space. As energy swings back and forth between the poles manifested by continuous vibration, it appears to rest at each pole before beginning the next swing. Thus, energy is said to be either in motion or at rest as it swings back and forth in an oscillating movement.

Back and forth.

Motion and rest.

As the swing between the poles increases in speed, the poles are said to draw back together until they converge into the whole that existed before they separated. But, conversely, when the speed of swing between the poles decreases, the poles are then said to separate and pull apart, creating more and more space as the distance of the swing widens and lengthens.

If energy did not oscillate, creation as we know it would not exist.

The observation and the study of this phenomenon is complicated, though, for according to science, you cannot see motion and rest at the same time even though they are aspects of the same basic sequence. You can see motion, as in the path a particle takes, or you can see rest, the particle itself in suspension, but you cannot observe the two at once, at least not scientifically.

In quantum physics, Heisenberg's uncertainty principle states that any attempt to observe the microscopic world can have an effect on what is being observed. Because of this, we cannot *prove* that absolute motion and absolute rest exist. Nor is there any way to know for certain if attempts to measure objects or incidents alter in any way what is being measured. Of significance here is the fact that what seems as certain—is actually uncertain.

And therein lies another paradox: our physical world appears as solid and stationary when it is anything but.

For instance, while you sit motionless in a chair, the molecules in your body are vibrating, all matter in your environment is vibrating, the earth is rotating about its axis while orbiting around the sun, and even the universe, as we understand it, is expanding. You think that by remaining still you are not moving, but that is not true. Motionlessness is filled with motion.

Only by separating the various aspects from the whole can we be certain of what we think exists; we can then study, examine, observe, analyze, and measure. But we can never measure simultaneously all aspects of the whole together, nor can we measure with certainty (thereby proving) what seems real to us. (As an example, a flash picture taken in a dark room does not show what the room was like while it was dark because the light of the flash made the room, for an instant, completely bright.)

Separation, then, enables us to be objective, but only the whole as a whole can help us to maintain perspective and context. What seems as whole is actually myriads of single units. Yet what seems as myriads of single units is but related parts of a connected whole. Neither can exist without the other, yet we cannot interact with both aspects simultaneously, (according to present-day science).

And therein lies the greatest of all illusions.

Because of the way our faculties operate and the way our brain processes information, we are conditioned to perceive everything as whole and solid when it is not. We create the reality we think exists by the way we connect together the data we receive within our own brain. What we see and hear and feel and touch and sense and taste and smell is totally and completely real to us, and appropriately so. But as near as science can tell, it is the length of the sine wave, that distance of oscillation between the two poles or points of rest, that enables much of creation as we experience it, to exist. This illusion of wholeness and solidity maintains its own integrity as long as vibrating energy oscillates rhythmically and nothing interferes with that oscillation.

Example 4: Quantum physicists tell us that everything that exists actually flashes in and out of existence about a billion times per second. First you see it, then you don't. During an "on" flash, existence is illuminated and everything is visible; during an "off" flash, there is only the darkness of invisibility and nothing can be seen. On and off. Back and forth. Motion and rest. Our built-in *perceptual prejudice* is what enables us to regard anything as continuous or solid. This natural prejudice shields us from the fact that motion and rest are separate sequences. We see solid objects and we see continuous movement and we think we are seeing both at the same time when, actually, we are not. The world around us exists as perceived because of how perceived. Creation, as we think it exists, is a physical illusion.

I have noticed that when vibrations within and around us speed up (and this can be sensed by anyone willing to do so), time is no longer able to act as a buffer between events that happen to us in the earthplane and our responses to them (that is, time whizzes by, there's never enough of it, the consequences of our actions manifest quicker). But when vibrations slow down, the span that exists between experience and thought (the "tit for tat" of cause and effect) widens and lengthens. Thus, the slower the speed of vibration, the greater the distance and the longer the timing between the events that happen to us and our response (that is, time pokes along, there's plenty to spare, we have all the time in the world).

To say this another way: Time protects the manifestation of existence space allows, so thought can reproduce itself.

11

Living in Time and Space Differently

When the heart weeps for what it has lost, the soul laughs for what it has found.
—Sufi Aphorism

Occasionally on our exploratory journey we need to address the whimsical caprice of human nature, for we cannot unravel all that future memory implies without recognizing what part this plays in the human family.

The transformation that comes when consciousness shifts, even temporarily, does indeed unveil greater truths about creation's story. Yet, that same enlightenment offers no panacea or quick fix to a better life. Those who experience such a shift eventually face this challenge: *Take the initiative to act upon the knowledge you have gained, don't just receive it; then accept responsibility for the power unleashed when you do.*

Two incidents that occurred shortly after my near-death experiences present us an opportunity to recognize what an individual can go through when the basis of his or her perceptual reality alters significantly. Realize as you read these stories of mine that awakening to expanded vision and greater truths is not some curious mental exercise. It is an intimately personal drama that can be as frightening as it is wonderful.

Internal Awakening: Reassessing the Perception of Self

After what happened to me in 1977, I could no longer relate to myself as Phyllis. Her habits, personality, clothing, possessions, and lifestyle were foreign and irritating to me; yet, when I looked in the mirror, there she would always be, looking back, all five feet seven, nearly two hundred pounds of her. She was shaped like a pudgy balloon with a deep sadness in her eyes and a washed-out face rimmed with short wavy hair. There was ample proof in her belongings that she had once been a bouncy, spontaneous woman with a love of song and a lust for adventure be it exploring caves or ghost towns, and she adored rocks. Everywhere I turned, there were rocks, even in her purse. Rocks in her head too, I mused. It took me a long time to figure out the whats and wherefores of my own existence, and I went through many stages in what later became a program of self-discovery. Reassessing the nature of "self" was one aspect of that program. Here is what I was finally able to clarify:

> I truly am an immortal soul, an extension of The Divine, who temporarily resides within a carbon-based form of electromagnetic pulsations that produces a solid-appearing, visual overleaf of behavior patterns more commonly referred to as "a personality." Phyllis is a name given to my personality, my temporal self, but the real me is I AM. And what I AM everyone else is, for all of us are cells in The Greater Body, expressions of The One God.

This clarification of my true identity flooded me with so much joy that I could hardly contain myself. Almost immediately, though, I ran afoul of a simple question: Since I am an immortal soul, why do I need a body?

The human body and my presence in one had to be faced if life on the earthplane was ever again going to make any sense to me. Now that I knew my greater identity, maybe my body had a greater identity, too. To find out, I took another look.

Examining my body more closely, I beheld a dense, slow-moving conglomerate composed of layers and levels of multitudinous life-forms arrayed in various sizes, shapes, colors, and designs. Each minute particle was quite alive, active and noisy, intelligent, a being unto itself. I was in charge of this collection in the sense that I as a soul provided the force field necessary to hold everything and everyone together and in place, similar to how God's Presence holds together and in place all of creation. Yet we coexist, my body collective and I, as teammates, a miniature version of the universe at large, a microcosm of the macrocosm.

This meant to me that I am the guardian of my part of the universe because mine is the major power source the others within me are plugged into. Should I leave, my teammates would eventually wither and die. As long as I stay, we can all then accomplish our given tasks, learn and experience whatever we need to, grow-change-recycle and grow-change-recycle, until we are free to disconnect our vibratory alignment and go our separate ways—die, if you will, to the physical

There is choice, mutual choice, theirs to be where they are, mine to be where I am. It is no accident we are together sharing the same journey, but it is mainly up to me to provide the direction we travel for I, as a soul, have a larger energy surge, am more highly developed, and can see farther ahead. As I am the source of power for the membership within my body, God is The Greater Source wherein my help comes. God, then, is the central current, that all-encompassing power source everything ultimately plugs into for the kind of sustenance that enables existence to exist.

In coming to grips with this understanding, life—mine and everyone else's—began to make sense. I could acknowledge my body as the living temple of The Living God, filled with beings cycling through births and deaths and rebirths as they, too, grew and evolved, all the myriad forms of intelligence, all the promise and the potential, all that was inside me . . . this my body, the incredible collective I wear.

What a privilege. What a remarkable privilege to be here, to be me, to be a soul sojourning through the earthplane with a whole conclave of intelligence accompanying me, a conclave that shares my challenges and opportunities unselfishly, dependably, lovingly. All of us evolving. All of us changing. All of us in the state of becoming more of ourselves.

How could I possibly ever abuse or misuse my body?

How could I reject it?

When I at last accepted my body, I also accepted the responsibility that came with wearing it. In recognizing that I was once again back inside its protective folds, I could then comfortably claim my body as "mine" and I could say the words, "I am Phyllis."

If you conclude from this rendering that I was completely detached from self and body as an integral unit following my near-death experiences, you would be correct. I knew myself to be a soul and I knew The Source Of My Being. What I could not relate to or even conceive of, however, was the idea that name-body-personality-incarnation could ever be separate or distinct in any manner from my true identity as a soul. This made no sense to me, neither did the earthplane.

When the day came that I could finally claim my persona and my body and feel good about that claim, I was surprised to discover how much my body had changed. There were differences, and those differences took some getting used to. Possibly my "new" body could have been explained as a by-product from the full regimen of health treatments I had committed myself to, treatments that included various natural healing disciplines and continual exercise beginning with the basics—things like crawling and learning the difference between left and right. Nonetheless, no matter how explained, there was no denying that my body had altered, and noticeably.

Yet, during the quest I embarked upon to seek out others as myself, I noticed that most other near-death survivors displayed virtually the same kinds of changes I did but without having taken the same steps I took to heal and readjust. This got me to wondering. After interviewing thousands of experiencers over several decades, I came to realize that the near-death phenomenon and other similar transformative events, if intense, complex, or at all impactual, can and do produce physical changes in the human body as well as stimulate mental, emotional, and spiritual awakenings or alterations. And adjusting to them can be an enormous challenge.

External Awakening: Reassessing the Perception of Time and Space

Before my transfer to Karen Woods's department at the Idaho First National Bank, a previous in-house move netted me a tiny cubical of an office in a corner location near the top floor of the old head office building. Huge windows filled two walls. There was no door, allowing my activities to be in full view of other department personnel. My boss at the time was Sandi Bonnett, a steady, unruffled type who went out of her way to be fair yet left no doubt who was in charge.

My main concern, though, was time. I didn't have enough of it. There was so much I wanted to do, so many things I wanted to think about, so many different places I wanted to be all at once, but there wasn't enough of me to go around nor did I have enough energy. Setbacks and disappointments had been many; my body unable to respond. No matter how hard I pushed or how hard I tried, I seemed to be going backward instead of forward, so I became obsessed with the notion of time. I wanted more of it, desperately, even if it meant restructuring my world and me with it. Perhaps this strong emotional desire caused what came next, I really don't know. I only know that whatever the cause, something utterly bizarre happened.

It was an exceptionally bright morning. Sunshine seemed to evaporate the windows as my office dissolved into light. Blinds were nonexistent, making the glare almost painful. Because of the brightness, I had to concentrate harder than usual at my drafting table as I prepared to put the finishing touches on a new bank form I was preparing for the printers. My eyes slipped for a moment past the edges of what I had drawn, and settled into a stare at the light blue background lines of measurement squares on the form paper I used. For reasons unknown to me, the blue lines seemed more important than what I had drawn, so I continued staring. As I did, each square spread apart. The blue lines bordering the squares elongated in size, buckled up from the paper, stood, stretched, then shot into the air near my nose. I had to jerk backwards to avoid being hit. The very second I did this, the expanding squares spurted forth what appeared to be clouds of energy that formed a veritable fountain of surging, glistening brilliance.

There was no way 1 could blame this incident on medication for I had taken none that day, so I started pinching myself to see if I might be daydreaming or perhaps suffering from eye strain. The pinches hurt. I looked out the doorway to see if anyone was watching, but everyone seemed quite busy and preoccupied. I squirmed around on my stool, then slid off to readjust its height. I glanced back at my drafting board as I did this, to see if the fountain was still there. It was, and was growing steadily larger.

As I put my hand to my mouth in disbelief and shook my head, a kind of vortex formed around the spraying fountain of energy, sucking in and pushing out masses of a mistlike, foamy plasma, until most of the board and the air in front of me filled with it. The misty plasma then began to narrow into the shape of rays, and from a central point in the middle of my drafting board, the rays spread open to form a fan, still growing and expanding in size until it reached halfway up the wall I faced. The glistening fan then divided itself into eight equal sections.

It was as beautiful as it was awesome, but before I could question the what or why of it, I felt energy from my own body move out and extend forth to join with the fan. This sensation continued until I could feel parts of myself existing within each of the eight sections. A protrusion from myself, like a stiff cord, lifted up from the back of my head and rose in the air about three or four feet. I again perched on the stool, for I was still me and was capable of continuing as myself. Quite suddenly, though, there were eight satellites of me in the fan and another extension of me above and slightly to the rear of my head. Counting the me sitting on the stool, there now was a total of ten divisions of myself, not just one—each fully alive as an independent mass while at the same time remaining connected to the others.

Once complete, the fan and the plasma dissolved into my satellites and each satellite then commenced to perform specific tasks, without any effort or thought on my part. One meditated. One prayed. Another continued an exploration of The Void. Still another went off to visit a friend who was ill. The four remaining engaged in various projects of thinking through ideas that were important to me. The extension from my head acted as a supervisor to make certain no satellite got lost or had any problems and that nothing was forgotten or missed. The original me remained in my original body and looked the same as always. The others, however, looked like sparkling masses of a mistlike energy.

At last there was more of me—to think, learn, ponder, explore, pray, investigate, meditate. What I had yearned for so desperately I now had.

Although no one else could see any of my extensions (I know, because I asked), I could clearly see, feel, and hear each; yet in no way did any of the satellites distract from what I was doing as Phyllis. What each experienced fed instantly into my own brain as if that brain of mine now functioned as a central receiving station. This situation lasted nonstop for ten full days and nights.

But there was a catch.

As time contracted into space during the incident, space expanded into time, then both converged. This created a state quite workable for short-term activity but not for long-term livability.

My physical body and all my physical movements slowed tremendously, like a forty-five rpm record being played at thirty-three-and-a-third. This meant that I could no longer walk, talk, or move in a normal manner but I could think and hear at remarkable speeds, and I could see as if my vision were multiangled. I experienced myself as superhuman, while, physically, I appeared subhuman.

To illustrate how absurd this was, let me describe what it was like to converse with people. I could see every minute motion in an individual's brain/mind assembly as it interacted to select words, then send those words through nerve channels until the result issued forth from his or her mouth. The size and shape of those words were plainly visible as thought-forms when each made an arc over to me, gliding along a "bridge" composed of the force from that individual's breath, saliva, and mental intent. These thought-forms entered my body as vibratory waves while registering in my ears as sound; then each moved through nerve channels into my own brain/mind assembly, where sparks and pulsations selected a response. And that response tracked back through the nerves to my mouth for projection, so it too could arc over, speeding along on a similar "bridge" to the person I was conversing with. Needless to say, a short conversation took forever. The process could be compared to several people attempting steady dialogue from opposite ends of an echo chamber. Driving, stepping up or down, lifting a knife to cut a potato, turning a dial, trying to squeeze toothpaste from a tube, tying shoe laces—each and every physical task required ridiculous lengths of time to perform.

Because of this physical slowdown, office personnel became frightened of me and complained to my boss. They claimed my office felt weird and I acted strange, almost as if I were caught inside a slow-motion movie. Complaints were so numerous that Sandi asked to meet with me. It took nearly an hour and a half to conduct our "brief" conversation. Sandi spoke of being patient and understanding with me, what with my health crisis and all, but whatever state I was in had to end because I was disrupting the entire department. I agreed to do what I could.

A sense of panic welled up inside me as I left Sandi's office, for I had no idea how to reverse the condition. But, since intense concentration and a strong desire based on desperation must have somehow initiated the event, I reasoned that an equally intense emotional focus could possibly reverse it. And that focus would not be at all difficult to create, since even though the last ten days had been a marvelous adventure, living life in slow motion was much too cumbersome for me and far too threatening for others. It took eighteen hours of deeply felt prayer before the reversal took hold.

Restoration to normal speed was as fascinating to experience as its alteration had been. However, midway through the process, I made a request. I did not want to return to full capacity. Slow motion did have some advantages in the way it enabled me to expand my abilities and thought processes, and in the way colors, sounds, and sensations were more vivid and the environment more fully textured. Although I had no further need for satellites or extensions, I did want to preserve what I could of the slower speed of vibration while still remaining functional and productive in society. I got what I wanted.

On the one hand, the time fan episode was immensely gratifying. After all, whether or not I understood how it happened, the fact is I had had a direct hand in altering time and space, literally existence itself. Somehow the very substance of the universe had turned inside out, then enlarged, and I had been there to witness the event and actively participate. As Lawrence LeShan had said, a reality is valid only if you can accomplish the goals acceptable to it, and I had done just that. Not only were my extensions completely real, the one that went to visit the friend who was ill impressed directly into my brain the current status of that individual, and this information was later verified when the friend telephoned.

On the other hand, though, the whole thing was troubling. That's because I had no control over the time fan's formation or how it affected me physically. Also, I began to feel like a freak when others complained with increasing frequency about how uncomfortable and threatening my presence had become to them. Since they could not share my reality, they were incapable of understanding it. This illustrated for me how defensive and self-protective people get if anyone or anything defies their sense of what is normal or acceptable (a parallel to the experiment with babies, as they were only startled when something violated common sense). The whole episode, though, taught me an incredible lesson about the responsibility I have to the collective whole and that I need to be more respectful of other people and their right to behave as they do.

In chapter 2, especially as concerns the photograph of the Grand Canyon, the image on Swedish television, and the herd of horses in Montana, the connecting thread that interweaves these three stories is strong emotional feelings and desires, and that is exactly what caused or is somehow responsible for the manifestation I had of the time fan.

This connecting thread leads us back to the limbic system, the seat of emotions and a passageway between what appears to be time and space so we can tackle the big one—existence.

12

The Innerworkings of Creation and Consciousness

Learning is the very essence of humility, learning from everything and from everybody. There is no hierarchy in learning. Authority denies learning and a follower will never learn.
—Krishnamurti

To tackle the subject of existence, we must confront the mystery of matter *and* the stirring of consciousness as it awakens unto itself, for the two, as you will see, are the twin building blocks of creation's story.

The word "matter," by the way, means "that which clings," indicating that matter as dense light explores, reaches out, and communicates with its environment until it congeals enough to cling together as a shape or evolve into forms. The way light solidifies into matter (congeals into substance) suggests that the universe is not so much based on a hierarchy of orders as on the collective evolution of forms and structures.

Yes, you read that right. Matter is dense light.

Exploring this, what matter is and how matter takes on physical shape, will enable us to probe many riddles, especially those of creation's inner workings and, interestingly enough, the evolution of consciousness as well. This exploration is necessary because the riddles we seek to solve lie at the heart of the limbic system and its passageway through "The In-Between."

To help us arrive at where riddles can be unraveled, we'll take a brief tour through the latest in mathematical and scientific discoveries about the mysterious world of matter. I assure you this tour will be worthwhile, even though it may seem a trifle confusing at first.

As velocity increases, time slows down. At the speed of light, time stands still. The faster something moves, the more energy it contains and the larger its relative mass becomes. (This exactly describes what happens to consciousness when freed from the limits of thinking during altered or flow states, perhaps even during a near-death experience.)

When the vibratory rates of light increase significantly (vibrate faster), time and space are revealed to be the illusions that they are. (The onset of a future memory episode displays these same conditions and characteristics.)

In Einstein's Theory of Relativity ($E = MC^2$), the speed of light is considered constant, matter and energy are interchangeable. Matter (energy) is known scientifically to "travel" in rings; light (the same basic energy as matter) supposedly moves in straight lines, yet is actually bent by the curvature of gravity and matter's unequal distribution. Matter is most commonly regarded in the field of science as solidified light, appearing as form and substance because of its density and lower frequency of vibration. Matter is not stable and only appears as such relative to how it is observed. Peter Russell, author of *A White Hole in Time*, offers this explanation: "Only as the Universe began to expand and cool were the elementary particles of matter created. In this respect matter may be thought of as 'crystallized' light; energy that has taken on the form and qualities of matter in the manner prescribed by Einstein's equation $E = MC^2$."[24]

Even though huge concentrations of energy are necessary to form matter, there is actually more energy in things other than matter than there is in matter itself. That's because matter and energy and light, at the most fundamental level, are the same thing.

Let me restate what I have just said.

Matter is dense light. And matter has the same duality as light. Light's duality is functional, behaving as continuous wave forms while at the same time appearing as discrete (separate and singular) particles.

No one can explain this, how light can be all things at the same time or even how light can "solidify" into matter once its vibration has slowed in velocity.

Space as a wavelength can increase, decrease, remain constant, or change, for we live in an open-ended universe that has no boundaries or edges. Time can reverse, stand still, or collapse as well. When this occurs, the illusion of time and space ceases to be relevant and the chaos of collapse reigns.

At a point of singularity, that point in space where everything compresses into itself and where the laws of motion and rest and location and movement cease, there is *convergence* (total collapse, a coming together— like what Itzhak Bentov referred to when he talked about how consciousness relates to time and space).

In the *Hammond Barnhart Dictionary of Science,* singularity is described as "surrounded by a region of space so distorted by gravity that nothing can escape; it is the region that constitutes the black hole."[25]

University of London Mathematics Professor John G. Taylor, in his international best-seller *Black Holes,* theorized that "Inside this object, the fundamental laws governing our universe appear to be destroyed, along with our usual concepts of time and space. The black hole not only puts the scientific world in turmoil but also challenges many of man's basic ideas about his surroundings and his place in them."[26]

Stephen Hawking, acclaimed English physicist and author of the phenomenal masterpiece *A Brief History of Time,*[27] took the reality of black holes another step by announcing that these so-called "vacuum cleaners of the universe" do indeed emit particles (radiation) and that the smaller ones can erode or explode. He established that black holes, large or small, are not separate and distinct from the universe but are, in fact, an integral part of the overall scheme of things and of the ongoing process of creation.

Certainly, black holes readily appear to be objects, where everything caught within one converges into total chaos. Theoretically, they constitute the ultimate singularity, the destroyer and death of existence as we know it. Yet black holes emit the very radiation suggestive of continuance and renewal. Whatever seems destroyed by them, or at least a portion of it, is somehow released fresh and new.

Could it be that black holes are connected to white holes? (Today's physicists conjecture that quasars are white holes, because the intense, superbright energy mass they emit is so extraordinary.)

Theory postulates that once everything compresses down through a black hole (implosion), part or all of it emerges back out through a white hole (explosion). As this is said to occur, radiation would naturally escape from the event horizon (the lip or mouth of a black hole). This theory, then, infers that creation has the ability to regenerate itself indefinitely through a continuous transmutation of matter/energy/light.

Black hole/white hole, convergence/emergence—the in/out flush.

Is this a mechanism for cosmic recycling?

Imagine for a moment that a black hole was stacked atop a white hole in such a manner that whatever was pulled into and down the black hole could then flush back out through the white hole beneath it. The resultant shape of the stacked holes would resemble an hourglass.

Is there anything in creation which could accommodate such a shape? Yes, the inside of a torus.

Whether energy swings in an oscillation or spins in a rotation, this process suggestive of cosmic recycling points to the presence and the importance of a special three-dimensional, geometric shape known as a torus, *the only self-organizing wave form known to exist.*

We need to know a lot about a torus, so I'll lay the groundwork here.

Mathematically speaking, a torus is generated by taking a sine wave (remember that the sine wave is the basic wave form in the universe) and rotating it on its own axis. The three-dimensional shape generated by this rotation resembles a doughnut in appearance. This torus "doughnut" owes its unique ability to self-organize and retain its own shape to three "gyroscopes" of spin inherent within it: up-down, in-out, and across. And, like a gyroscope, no matter how the torus moves, it maintains its own balance. An example of this in our everyday world would be that of a smoke ring. A smoke ring is capable of remaining intact as it moves across a room because it is a true torus.

Because a torus is so basic and so dependable in its ability to self-organize, its structure may explain how black holes and white holes could possibly recycle universal substance. In fact, both holes may simply be opposite ends of a sizable torus. With this in mind, let's investigate the idea.

Since a torus looks like a doughnut, envision a doughnut—an unusually large one:

Drawing used with permission from Vital Signs *Magazine, vol. 4, nos. 1 & 2, Summer/Fall 1984 (a publication of the International Association for Near-Death Studies). Refer to Resource Section for particulars about IANDS.*

Then visualize that as existence evolves, matter/energy/light moves along a curving pathway similar to the outer circumference of a torus doughnut, irresistibly drawn as it travels to the gaping mouth of the doughnut's swirling hole. This action most probably duplicates how water is pulled into and down a sink drain (follow the upward moving arrows in the diagram to the visible hole at the top). Once existence compresses into the hole (converges), this same energy could later emerge from the other end (the bottom hole) rejuvenated, renewed, and ready for another trip up the curve. Each revolution up that doughnut curve and through the "cosmic drain" (black hole/white hole) would cause a slight step-up or increase in the overall vibration of the doughnut. If this analogy holds, then the process just described would imply that as existence evolves, *so, too, could that which contains it*!

As you look at the diagram of the torus doughnut, doesn't it also remind you of a bubble? What if creation, as we think it exists, is contained within "giant bubbles"? The torus does supply the proper shape, for such bubbles would need to be somewhat flat on top and bottom (as is a torus). There is precedent for size, as energy envelops and circulates around our planet in the exact shape of a torus (with the north and south poles being the torus "holes").

Could it be that the current notion of the universe consisting only of flat disks called galaxies spinning in a vast nothingness is erroneous? What if the entire universe is one huge torus, or a bubble the shape of a torus, or a torus within a torus within tori? The prospect of a bubble universe, whether based on the torus concept or something else, is not as far-fetched as it may seem.

J. Richard Gott, an astrophysicist at Princeton, has proposed that the universe we live in is but one of an infinite number that were created, like bubbles in a hot liquid of intense but finite density. His calculations suggest that each of these universes is infinite or "open" in terms of its potential for expansion. The extremes of density and heat necessary to have resulted in such bubbles would have meant that at one time gravity was unified with the three basic subatomic forces (strong nuclear force, weak nuclear force, and electromagnetism). If Gott is correct, this means that a unified field theory is not only possible, it is an absolute necessity if we are ever to understand the uncertainty of that which seems so certain.

Again, matter always travels in rings.

A torus is a self-organizing, ringlike shape, not limited in size (microscopic or gigantic), which supports and maintains matter as matter travels in rings.

The integrity matter seems to display comes from the gyroscopic ability of a torus to balance and steady itself to appear "solid."

This we know.

But what if, in the middle of a torus, forces and forms do indeed converge, transmute, and recycle, so matter in the form we are familiar with can be continued . . . as it travels in rings as rings?

In and out and around.

Back and forth and throughout.

This idea reminds me of breathing, as if our universe and that which contains it, plus all other existences, seen or unseen, were alive and growing. There is a pulse, a beat, rhythm, order.

Regardless of what seems to be chaos, order reasserts itself sooner or later. Always. (This fact is what allows a laser beam to exist, for the singular intensity of a laser ray results from the combined power of molecules oscillating at thousands of different frequencies within it. The very chaos of this diverse mix is what produces the laser's incredible coherence.)

As is true with holograms, any part of a whole can reveal the whole. Our universe operates like that, like a hologram. So when you slice an apple in half (part of the whole), you see the basic design of a torus doughnut (the whole itself); or when you attach electrical probes into both ends of an egg or a seed and watch the flow of current between the probes, a torus takes shape, although more elongated like an ovoid; or when a fertilized egg becomes a fetus and the contents therein turn inside out to form a torus.

Obviously, the torus shape is the basic energy-flow design the pulse of life favors. Whether the idea of black hole/white hole in-out flush is correct or not, the question still remains: what really happens in the middle of a torus where forces converge?

Perhaps this puzzling riddle about convergence can best be approached by recognizing what happens when water is stirred.

There are several ways you can stir water. For the sake of this discussion, let's rotate it. Spin the water. Round and round. Faster and faster. Then stop the direction of the spin. Stop it dead in the water, and reverse the direction.

Did you see what happened?

When you stopped the spin, the water collapsed into itself creating an implosion. But just before you initiated a reverse spin, where the water could explode back out again, conditions mysteriously changed and both the water and everything contained within the water were briefly held in suspension. This is called a colloidal condition and particles caught therein are referred to as colloids.

As we needed to know a lot about the torus, we also need to know as much as we can about the colloidal condition and colloids.

(I promised you this brief tour of scientific ideas and discoveries would be worthwhile, and it will, so hang on a little longer.)

Routinely, science considers any substance that does not completely dissolve but that remains suspended in either gas, liquid, or solid, a colloid. Examples of colloids are rubber, cellulose, starch, protein, plastic, and nylon. No matter what you do to them or where you put them, these substances do not sink or disappear or blend, but remain suspended. They exist as themselves, by themselves, indefinitely.

I believe the colloidal condition describes what happens after everything converges inside the doughnut hole of the torus. This colloidal condition reflects the core mechanism for the shift energy makes, a shift that suspends, expands, and transmutes whatever is caught within it, either instantaneously or seemingly so.

To gain a better grasp of what I am saying, look again at that water in a colloidal state of suspension.

Right after the vortex of spin collapsed, surface tension increased dramatically; then antigravity ensued (similar to what is thought to happen once inside the event horizon of a black hole). And anti-gravity continues to exist quite apart and disconnected from the water's movement until the reverse direction of rotation can be generated (or until the condition dissipates because of no further movement).

Don't confuse what is happening here with lack of gravity, a weightless state experienced by astronauts in outer space. The colloidal state and the weightless state are not the same. Weightlessness is really a "free fall," like the slight sensation you feel when an elevator stops suddenly yet it seems as if you are still moving. Astronauts train to experience prespace weightlessness by briefly floating around in an airplane cabin as the plane dives earthward at a speed equal to the acceleration of gravity.

The colloidal state is a peculiar in-between condition which results when forces suddenly collapse, then converge. This in-between state creates antiforce, which is antigravity. Particles caught in this unique state between implosion and explosion transmute, and remain forever changed by that transmutation. On a molecular level, these particles show evidence of enlargement and of having taken on different and enhanced characteristics.

It is my belief that if we could understand the full import of what happens in colloidal states and of the antigravity created by them, we could uncover the secret to interdimensional and intradimensional travel, stellar travel, and to whatever The Void might be. Not only that, if we could further explore these states, I believe we could uncover how the function of a torus might indeed be the model for the transformation and transmutation of energy as it cycles through various stages the equal of creation itself.

Regardless of whether or not this model continues to prove itself as science becomes more sophisticated, the concept it offers of how the various forms of matter/energy/light could evolve by being recycled remains persuasive.

In review:

Black Hole: Implosion, the destruction of form, death (metamorphosis).
White Hole: Explosion, the manifestation of form, birth (evolution).
In-Between: Colloidal state, suspension/expansion/transmutation, unlimited potential (The Void).

I mentioned gravity.

If you really study what you can about gravity, then to the best of your ability test what you learn, I think you may notice what I have: gravity does not behave as a force; it behaves as if it were the entrainment of the spin of a smaller object by a larger one—literally the attraction matter has for itself. When such a spin is suddenly reversed, the entrainment (attraction/coherence/resonance) momentarily collapses until the reverse spin can begin. During the collapse, that which was held together by mutual attraction is freed. To put it another way, when the entrainment of relationships-in-motion-spinning-together suddenly stops, suspension results and antigravity ensues (a colloidal condition).

The same thing can happen to the human brain if suddenly hit, jarred, or severely jiggled, especially during an automobile accident or as a result of a fall. Typically a colloidlike suspension of consciousness will follow such trauma, wherein environmental space appears to expand out as time slows to a standstill. The individual feels somehow caught in between realities when this occurs, as if he or she had slipped through a crack in time and space and had suddenly become resident in a world "neither here nor there." This peculiar "feeling" of being suspended in between realities makes such an impression that it can permanently alter the way the individual regards the world at large and his or her place in it.

Of interest here is that consciousness, even when simply released from the bias of thought (as in a flow state), will behave in a fashion similar to gravity when gravity is freed from the attraction/entrainment that seems to have both caused and maintained it. This similarity demonstrates that the same type of collapse followed by a state of suspension, no matter the cause, can lead to sudden mental enhancements that can liberate the individual's potential (thanks to the emergence of antigravity, or antiforce).

Flow state similarities are not the only comparison that can be made here. You can also compare modes of awareness and their brain support structures with the idea of black hole/white hole/in-between. You can do this by utilizing the language of symbology.

Traditionally, the color black is equated with subjectivity, the feminine, that which collects, absorbs, intuits—the realm of the subconscious, where everything dissolves into the darkness of invisibility. Conversely, the color white symbolizes objectivity, the masculine, that which analyzes, clarifies, separates—the realm of the conscious, where everything stretches forth into brightness and is readily seen. The In-Between is considered that which contains and embraces and unifies all—the realm of the superconscious, where everything becomes more of itself as it converges back into the collective whole. Devoid of any specific coloration, the In-Between is said to be radiantly luminous.

If you combine a shortened version of the chart on "Correlations within the Brain/Mind Assembly" (found in chapter 9) with a short summary of our present discussion, a provocative comparison emerges (see "Inner Workings of Creation and Consciousness" chart).

In the manner that creation seems to transform and transmute itself, so too does consciousness. Experience has shown me that both utilize the same fundamental systems to accomplish this task. Whether through a black hole/white hole in-out flush (or something similar) or through a transformation of consciousness (no matter how caused), the need is the

THE INNER WORKINGS OF CREATION AND CONSCIOUSNESS	
CREATION	**CONSCIOUSNESS**
Black Hole: Implosion—the destruction of form; death (metamorphosis)	Right Brain: Subconscious/ subjective—absorbs/enhances/ connects; imagination and intuition (religion)
White Hole: Explosion—the manifestation of form; birth (evolution)	Left Brain: Conscious/ objective—analyzes/clarifies/ separates; intellect and education (science)
In-Between: Colloidal state—suspension/ expansion/transmutation; unlimited potential (The Void)	Limbic System: Superconscious gateway/ collective whole—senses/ embraces/unifies; mystical knowing (gnosis)

same, the process is the same, the goal is the same—to release old forms as new ones emerge so that which exists can expand and grow. Once again, only this time step-by-step, a colloidal condition is where

- forces suddenly collapse, then converge

- a momentary state of suspension results

- everything caught in that suspension expands and enlarges as antigravity is created

- inherent or unlimited potential is released

- whatever is present is imprinted (becomes permanently altered by what happened)

- whatever is present then transmutes (takes on different characteristics)

- as reversal of motion is completed, forces are restored, suspension ends, but the imprinting (transmutation) remains.

I believe that the common denominator linking consciousness with *creation—the single lynchpin that governs the alteration and evolution of consciousness—is a colloidal condition,* not just within a single individual's mind as that person matures or changes, but collectively through the combined effect of human consciousness as a whole.

In chapter 8 we had an opportunity to explore several types of flow states, all made possible because of our receptivity to them and our willingness to surrender to "no-thing"—a peculiar realm beyond self, where no one or nothing seems in charge yet unlimited potential awaits. Past, present, and future time spans collapse in this nothing place and *everywhen* reigns (as poet Gary Snyder would say). One returns from such a flow state quickened in a refreshing way and possessing more knowledge than before.

It has been my observation that this tranquil or slower shift in consciousness leads to milder versions of the colloidal condition, because that's what a flow state is, "a crack in consciousness," a temporary respite leading to the in-between worlds of colloidal environments. Meditation, spiritual and religious disciplines, and various traditions of the shaman, or mystic, are all designed to guide one to this very different place inside one's own mind. Thus, the colloidal condition can be taught and cultivated, either through continuous practice or by using certain rituals or techniques.

The turbulent approach to a consciousness shift, that is to say a spiritual breakthrough, is more rigorous, even violent, as forces within are released so fast immediate convergence results. Traditionally, any danger to the individual was said to be minimized by the attentiveness and preparatory instruction of experienced teachers or elders. This accelerated, or faster, way was and still is actively courted, for it enables one to penetrate quickly into full-blown colloidal states. This sudden, sometimes forced entry can foster powerful, longer-lasting illuminations of spiritual enlightenment; but the turbulent approach is unpredictable, even under ideal conditions, unpredictable in the sense of whether or not it ever happens or how the individual might be affected when it does.

Amazingly, this same immersion into a world in between realities can happen to people who are totally unprepared, unaware, uninformed, and uninterested—and this exactly describes the universal experience of the near-death phenomenon.

People who go through such transformational events become more of themselves because the potential inherent within them expands when released (the same thing happens with colloids).

Especially in near-death cases, you hear experiencers admit that what happened to them was what they needed, that it was somehow "ordered." And, if you keep that person's near-death scenario in context with the life he or she lived, you see connections, correlations, parallels. It is as if the phenomenon is one of nature's more accelerated growth events—a powerful dynamic that can foster the integration of body-mind with spirit while keeping that particular energy mass (the individual's soul) on track within its own evolutionary spiral.

I identified four distinctive types of near-death experiences in my book *Beyond the Light: What Isn't Being Said about the Near-Death Experience.*[28] Recognizing that these experience types do match, at least in a general way, the growth potential unleashed in the one who undergoes that particular episode, I offer this synopsis as a brief rendering of the four types:

Initial Experience: An introduction for the individual to other ways of perceiving reality; stimulus.

Unpleasant and/or Hell-like Experience: A confrontation with distortions in one's own attitudes and beliefs; healing.

Pleasant and/or Heaven-like Experience: A realization of how important life is and how every effort that one makes counts; validation.

Transcendent Experience: An encounter with Oneness and the collective whole of humankind; enlightenment.

If you are objective about this synopsis, what emerges is a fascinating panorama suggestive of what could be the natural movement of consciousness as it evolves through the human condition via stages of awakening. These stages of awakening extend from the first stirring of something greater, an initial awareness, to confrontations with the bias of perception followed by opportunities to cleanse and start anew; then leads to the bliss and the ecstasy of self-validation and the discovery of life's worth until at last the moment comes when unlimited realms of truth and wisdom are unveiled.

This panorama of awakening consciousness indicates to me that the near-death phenomenon is more than just a singular anomaly occurring to only certain individuals. More importantly, it seems to be *part of an ongoing process of transformative adaptation within the human family as a species*, a particular growth process that appears to shift individual souls from one stage of vibratory awareness to another or from one state of embodiment to another, a process that parallels and duplicates what happens during a spiritual transformation.

Remember, a colloid is any particle caught in a colloidal condition. Once suspended in that manner, the particle will automatically enlarge and expand and remain permanently and forever altered by the experience.

The process that creates a colloid correlates to an incredible degree with what happens to those who undergo spiritual transformations and near-death experiences. The majority who go through such a process experience an enlargement and expansion of consciousness, exhibit the sudden surfacing of latent abilities, face a confusing array of psychological and physiological aftereffects, and are never quite the same again.

It is my belief that the reason this process of convergence and transmutation (transfiguration) is so universal and basically the same for all is because all of us, everyone of us, now and throughout all ages, are and have always been imprinted by the same creative impulse that originated us. The mark of our creation is what we display whenever our consciousness is freed to rediscover itself and the source of its being. After such transformative experiences, we feel as if we have found "home" because the home we think we've found already exists within and always has.

We recognize the place because we never left it to begin with.

Separation, what makes us think we have left our source, is really but a slip of memory, a case of mistaken identity. We make this mistake by identifying with the projection we have become rather than with The Source of who we are.

The only purpose of a transformative experience, in my opinion, is to help us remember what we forgot.

Consider this:

- The way to access pure consciousness is through a flow state.

- The way to evolve embodied consciousness is through time-space ego states.

- The way to transform embodied consciousness is through colloidal states.

Whatever we need to awaken to the truth of our being appears to manifest when we need it most, and the way this happens is basically the same for all because we are all part of the same consciousness.

I have come to realize that the near-death phenomenon and spiritual transformations are rites of passage through the labyrinth of mind and matter into the light truth brings. In these rites of passage, experiencers are propelled by the power of a colloidal shift from one realm of awareness to another, switching realities en route while expanding into vision. Perceptual preferences dissolve or change when this new light, this knowing, dawns.

Previously I asked you to remember the significance of the limbic system in the brain. Here's why. By tying together what has been said about the limbic with the comparison chart in this chapter (entitled "The Inner Workings of Creation and Consciousness"), the puzzling riddle of convergence begins to unravel.

Here's the full riddle: If the limbic system is excited or stimulated enough to activate states other than its normal "housekeeping" directives and maintenance chores, is it possible to slip through the limbic gateway, through the "zero point" it offers, into "The In-Between"?

Here's how the riddle unravels: the answer is yes, because "The In-Between" literally means *in between* the on/off switching of brain neurons, *in between* the on/off flashing of physical manifestation, *in between* all vibrations and movements of energy and the eternal pattern of motion and rest, *in between* the countless variations of interdimensional and intradimensional realities and parallel and alternate universes *all caused by and resultant from the wake of energy's need to vibrate, in between* all the phenomena of existence and nonexistence, life and death, to the heart of gnosis (knowing) where the collective resides—that shimmeringly luminous potential of The Void—and even beyond The Void to God Itself, whatever God may be, for it is through The In-Between that we eventually encounter the purity and the power of indivisible consciousness.

Remember what happened in Bentov's diagram of subjective and objective time and space, especially when subjective space overlapped objective time and objective space overlapped subjective time? This created the same type of convergence that we have been seeking to understand. Whatever is caught in such a condition not only converges but suspends and expands as well. When this happens, normal coordinates and definitions disappear, limitations dissolve; everything can be anywhere instantly. As Bentov said, "We can fill the entire universe with our consciousness at infinite speed and be everywhere at once, in other words, become omnipresent."

What we can see in the microcosm with the stirring of water, we can recognize in the macrocosm with the stirring of consciousness.

13

The In-Between

No theory of physics that deals only with physics will ever explain phys-
ics. I believe that as we go on trying to understand the universe, we are
at the same time trying to understand man.
—John A. Wheeler, Ph.D.

I have two stories to tell you: one involves an encounter with pickle syrup that occurred when I was a young mother and the other is from my third near-death experience, when I was confronted by two cyclones. I know these seem unlikely topics, especially now, but, bear with me. Each will edge us closer to "The Dimension of Patience" located at the centerpoint of the labyrinth we travel.

The pickle syrup incident occurred after I had joined an Edgar Cayce "Search For God" study group in the midsixties. Being a member of this group was important for me, as this was my first introduction to a structured and in-depth examination of metaphysics (that which is "First Cause," the ultimate nature of existence), and to the realization that the so-called "paranormal" is actually quite commonplace, a natural adjunct to healthy living. (Edgar Cayce, considered by some to be the greatest psychic who ever lived, died in 1945, leaving behind a legacy of records, transcripts, and documents unequaled in the literature of parapsychology).[29]

Study Group Two in Boise, Idaho (the one I joined), was a lively and enthusiastic conglomeration of people. It was not unusual for us to meet two or three times in any given week and pursue homework assignments as well. We were voracious in our hunger to learn whatever we could about God, Truth, Life.

One discourse from the Cayce collection bothered me greatly, though. While in a trance state, he was asked to name the three dimensions governing the earthplane and he replied, "Time, space, and patience." He then went on to describe patience as not just a virtue or attribute, but as a dimension of existence, a physical dimension, something beyond time and space yet auxiliary to both. This made no sense to me.

One evening, however, while I was adding the last increment of sugar to a batch of fifteen-day sweet pickles that I was canning, a most horrific thing happened. The huge kettle I used to prepare the pickle syrup, a kettle nearly filled with thick bubbly-hot liquid, boiled over. I mean all over. Imagine if you will the scene that followed, for that thick sugary goo not only spread out to cover the entire stove top but it quickly coated the inside surfaces of all burners, drip pans, oven, inner framework of the whole appliance, the large drawer beneath the stove, all pans in that drawer, the sides of nearby cupboards, until finally puddling underneath and to one side of the stove.

The mess was unbelievable. Yet, what happened next was even more unbelievable.

I was at the kitchen table readying jars when the syrup boiled over. Instead of screaming in horror, which would have been normal for me, I calmly stepped forward into what seemed to be a resplendent, gauzelike, misty netting. As I did this, time and space seemed to overlap, fold, then converge into each other, while my movements slowed tremendously. The integrity of my kitchen as it existed, even the very fabric of the air, shifted. Dimensions switched.

When I reached for the kettle, it was as if my arm glided through the bright gauze, which I experienced as threaded strands of tiny, sparkling bubbles—microscopic, but clearly visible. My face, my entire body, registered sensations of touching and being touched by this stringy fabric netting. There was a smell present similar to that of ozone with a slight hint of ammonia, not pungent but "flat."

Without thought or effort, I slowly, gently, and ever so easily removed the kettle and put it in one section of my divided sink, turned off the stove, filled the other section of the sink with hot sudsy water, and proceeded to wash every single inch of every single thing. Then I unplugged the stove from the wall and pulled it out to the middle of the kitchen, mopped the floor, pushed the stove back, and replugged it. Not just once did I scrub and mop, but three times. Sugar syrup is quite sticky, you know.

When the job was completed, I looked at the clock. Six minutes had elapsed. *Only six minutes*! At least forty minutes should have passed, possibly an hour, for the amount of cleaning I had engaged in was extensive.

The surprise of seeing the clock snapped me out of the state I was in. Dimensions readjusted, while time, space, motion resumed their rightful speed and proportions. There was enough syrup left that, with added sugar and water, I went ahead and completed the project. It wouldn't taste right but at least the job would be done. After finishing, I sat down for a while, my mind racing with questions.

When I had first stepped forward, it truly seemed as if I had stepped from one dimension of existence into another. Even the composition of the air had changed to that of a buoyantly touchable, completely visible substance of a stringlike netting. Colors, sounds, sights, sensations, even smells had somehow been enhanced and altered.

When time and space had ceased to exist as the states they once were (what I was accustomed to), unlimited versions of both had instantly filled my kitchen, enveloping me as they did. This peculiar paradox had enabled me to access infinite amounts of pure energy automatically. I had become a human dynamo because of this, accomplishing the impossible without the slightest effort or tension and in less time than imaginable, yet my awareness of what was happening as it was happening, and my actions and activities throughout, had been that of total *slow motion*. No decisions or emotions had been involved. As I had gone about the task of cleaning up the mess, everything, including me, had felt weightless, bouncy, peaceful, and harmonious. And I had felt unified with everything, as if I and all the objects in my kitchen were connected to each other and similar in essence.

Was this what Cayce meant, I wondered. Had I accidently discovered the dimension of patience? Had I discovered the dimension where effort and stress were nonexistent, energy unlimited and readily available, and time and space the same as nothing at all?

This incident profoundly affected me and inspired further search as to what "patience" might be. Over a decade later two comparable incidents occurred—the cyclones and the time fan. Since I've already discussed the time fan, I'll tell you about the cyclones.

Of the three near-death episodes I experienced in 1977, the third one was the most dramatic. And it haunted me. It intruded upon my life, becoming more detailed and more powerfully real as years passed. It would not leave me alone. The actual scenario had involved huge energy masses the shape of two cyclones, one inverted over the other. I initially described them in chapter 2 of *Coming Back to Life*, but that description was not complete. I left out some of it. Here is a more detailed version:

During the evening hours of March 29, 1977, when I left my body in what felt to be death, I soared rapidly through the roof of the house I rented, glimpsing each molecule of material in the ceiling and rooftop as I went and noting how curious it was to possess such X-ray vision. As if flying, I rose far into the night sky until deep in heaven's darkness I spied a slit of brilliant light somewhat the shape of a "lip." When I neared, the lip of light opened slightly, enough to allow entry, but that entry was more an absorption, as if I had suddenly become caught in a force field. This "field" extended some distance into space and away from the lip. Particles of twinkling brightness identified its presence. I detected the smell of ozone, increasingly "flat" as an odor the closer I got. Once inside, the light was as overwhelming as it was brilliant, yet it had no apparent source. I spied two colossal forms in the distance, cyclonic vortexes spinning at great speed, with one inverted over the other in an hourglass shape.

The cyclone on top spun clockwise. The inverted cyclone beneath spun counterclockwise. Where the two spouts should have touched but didn't, there spewed forth in all directions piercing rays of radiant power—not light, power. Power!

Both cyclones were fat and bulgy, not at all smooth-sided as might be supposed, considering their tremendous rate of spin. Even though the direction of their movement was decidedly right to left for the one on the top and left to right for the one on the bottom, inside each was the presence of the other's motion plus a separate inner convolution. This tridirectional force seemed to create the powerful spin along with an impression of layering across the surface of the cyclones (without rows or bands to cause the layered effect).

Inside the top cyclone (and I called them cyclones because that is what they reminded me of), I saw my Phyllis-self, hardly larger than a speck, yet recognizable. Superimposed over my Phyllis-self were all my past lives and all my future lives happening at the same time in the same space as my present life. Around me were other people I knew. The same thing was happening to them. Around them were still other people, and others more, until I came to realize all life-forms were present inside the cyclone, and the same thing was happening to each and all. Yet no one and nothing made any "real" movement except expansion and contraction, as if all life, plus the environment in which it existed, was breathing.

What appeared to be movement, the life-forms acting out their given roles, was actually an optical and perceptual illusion, similar to a hologram but produced by pulsed wave oscillations activated by individual and collective forms of consciousness. If any life-form changed the overall pattern of a personal scenario, "past" as well as "future" would alter for that individual and sometimes for others. While each life-form was truly its own self, each was also connected to all others by bubbly threads of a brilliant light that formed a fabric netting or web.

And what occurred inside the top cyclone also occurred inside the bottom cyclone. As above, so below. In other words, my Phyllis-self plus the other life-forms actually inhabited both cyclones in the same relation, in the same condition. The bottom cyclone, then, was but a mirror image of the one on top. The overall scene first impressed me as if a giant echo were filling the width of a massive canyon.

The sheer force of cyclonic spin created a counteractivity along each of the cyclone's outer edges, manifesting in the process another energy construct altogether. This extra construct occupied space to the left and to the right of the cyclones and seemed somehow to originate darkness and light as by-products of its existence; thus, darkness developed to the left as light emerged from the right.

This sight filled me with the realization that darkness and light, by-products of the spinning cyclones, were opposite "signatures" of the same dynamic. They provided the necessary mechanism and contrast for manifestation to be experienced in a meaningful way. Darkness and light, then, were corollary reflections resulting from the act of creation continuously re-creating and altering itself, for that is exactly what it felt like, as if I were witnessing Creation.

Since what I had once referred to as "life" no longer interested me, I found myself fascinated with the rays of radiant power, those piercing rays continuously emanating from the middle, where the cyclone spouts should have touched but didn't. That space, that place, seemed to me as if it were the entry way to God, so I resolved to go there, to head directly for the centerpoint. My desire was to return to the God from whence I had come. God!

At that moment back in Boise, my son Kelly found my body in the living room and began to talk to it, speeding words my way, tones, and I heard him. I have no memory of what he said since only his tone mattered, for riding on his tone came love, unconditional and freely given. That caught my attention and turned me away from the radiant rays. Had I made it to the middle, there would have been no coming back. I knew that. But to the middle I almost went before the sound love makes reversed my direction.

If Kelly had used the telephone to summon emergency assistance after he found my body, needed help would have arrived too late. Fortunately, he obeyed what he had been taught to do since his youth, and that was to always consult the inner self (internal source of guidance) before contacting others (external source of assistance); go within before going without. This inner wisdom inspired him to pull up a chair opposite my body and start talking, no special words, just nonstop sound to create an audible current. This he did and I responded. I glided back on that current of his sound, back to my body and to my life as Phyllis.

Obviously, I am not an artist, but take a look at this sketch I made of the cyclones and the side activity that produced darkness and light (see "Cyclone 1" on the next page). The rays were impossible to draw so I simply extended lines with a yellow marker; unfortunately, the color disappears in printing so I used heavy ink over the yellow; What seems to be a white spot in the middle is not meant to depict a hole but, rather, the place from which the power rays came (that combination of forces feeding into the middle of the cyclones combined with forces central to that location).

Now, take this same sketch and turn it to a vertical position instead of horizontal, with darkness at the top (see "Cyclone 2" on the next page). Does the scene remind you of anything in particular? Could it possibly be a cutaway drawing of a torus? The three spins are there. Anything else?

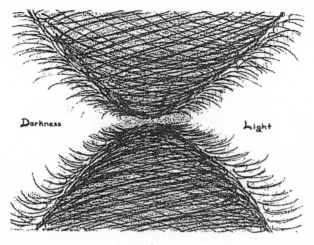

Cyclone 1

Use your imagination here. Envision the darkness at the top as a black hole, the light at the bottom as a white hole, and the middle, where the rays are, as the point of convergence. Then imagine that this whole construct actually gives form to a process as well as an object, an ongoing process of creation recycling itself.

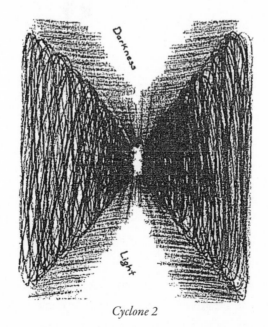

Cyclone 2

Here is another version I drew of the same scene, as I sought to remember as many details as possible of what I had witnessed:

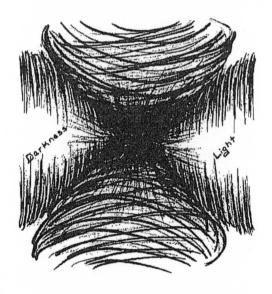

To aid your imagination, here is a cutaway drawing of a torus. Turn the book around with your hand and view the drawing on this page from several angles, then compare it to my sketches of the two cyclones, both horizontally and vertically. Can you see a similarity?

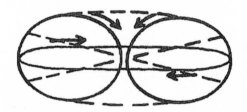

Drawing used with permission from Vital Signs *Magazine, Vol, nos. 1 & 2, Summer/Fall 1984.*

More was involved in what I saw than has been pictured. Particles of twinkling light extended some distance from the lip of light, creating a force field. Also, the activity to either side of the cyclones seemed indicative to me of yet another construct, that of a system within a system.

Appreciating that what I saw may indeed have been the middle of a torus, as I believe it was, then this side activity exposed the presence of another torus, one inside the other. When I pulled back to hear my son's voice better, I took one last look at this scene and beheld as I did a panorama so awesome it haunted me for years afterward.

Not long ago, while awaiting my flight at an airplane terminal, I found myself absentmindedly flipping through a paperback I had just purchased entitled *Steven Hawking's Universe: An Introduction to the Most Remarkable Scientist of Our Time*, by John Boslough.[30] Midway through the book was a photo of Hawking's collaborator, Roger Penrose, looking at a drawing on a blackboard of virtually the same thing I had previously sketched of that final glance before leaving the realm of brilliance and reviving back inside my body. The only appreciable difference between the Penrose version and my own is that mine has a funnel shape extending from each end of the torus-within-a-torus, while his featured a long narrow tube. My sketch with the funnel shapes at each "hole" is shown below in partial cutaway:

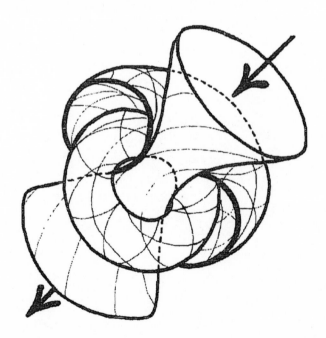

The caption under the Penrose photograph states that the blackboard drawing is an illustration of how time might be produced (presumably once matter/energy/light flushes through the hole in the torus doughnut). The radiation (particles) Hawking placed around the event horizon or lip of the hole is represented as the extended passageway protruding from both ends of the larger torus. The second torus, the smaller one inside the larger one, illustrates, at least to me, that several systems are at work here to ensure the continuance of creation as we know it. I cannot duplicate the mathematics used to make the blackboard drawing in the photo, but apparently, without any intention on my part, I did duplicate the picture.

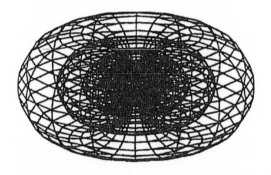

Toroidal fields, nested one inside another with
concentric vortex throats

Energy donuts in symmetry

Computer simulations of a torus-within-a-torus.
Graphics from Dan Winter and Friends, Crystal Hill Multi-Media

Future Memory

What I actually saw and felt during my third near-death experience and how I finally captured that scene on paper closely matches the physics of time/space/matter, plus the theory of creation. I claim no expertise here, but I do know what I encountered and it was *very, very real.*

Years after my near-death experiences were over, 1 discovered that history and legend are filled with reports of people who, having had impactual transformations either because of nearly dying (usually a near-death experience) or from a total change in consciousness (usually spiritual enlightenment), described something akin to what I saw—a shape the likes of the cyclones. Gifted psychics have spoken of the same thing and so have people on their deathbed as they were about to die. In fact, *a large, predominant vertical shape* such as a column, stairs, beam of light, great tree, or hourglass image of vortexes (similar to the torus "throat" I saw), is *the most repetitive motif* found throughout the whole of visionary symbology. According to tradition, for one to have witnessed or traversed "The Vertical" [see appendix IV] is considered a sign that the individual has transcended the "twelve heavens and twelve hells" horizontal to earth's vibration, and is ready to move on to higher realms of grandeur beyond what the human mind can fathom.

Since to have seen the likes of what I saw, to have experienced it, is so much a part of transformational literature and lore, what then is really going on here?

I think that what happens to people like myself is that we inadvertently or intentionally enter the passageway into the colloidal condition and experience in so doing an "otherworld" journey that is actually the colloidal state itself (regardless of what form, image, or scene is initially witnessed). Having reached this level of "truth," we are undeniably marked in the way we change afterward.

Black hole/white hole/in-between; right brain/left brain/limbic system—creation/consciousness—is not one the reflection of the other? Is not the centerpoint the goal of each of us? Is not the centerpoint we seek located within and through the colloidal state of convergence that we enter via the limbic gateway?

Allow me to repeat:

External Colloidal Condition: That state of temporary suspension that results from the collapse and compression of time-space states and relationships-in-motion-moving-together. This releases unlimited potentiality through the sudden expansion and enlargement of energy/matter as its form and substance are altered.

Internal Colloidal Condition: That state of convergence that enables subjective and objective states of consciousness to momentarily overlap and expand, thus making it possible through consciousness to be anywhere and everywhere simultaneously and with a marked acceleration of awareness and knowing.

Colloid: A particle or potentiality caught in a colloidal state that is enlarged, enhanced, and expanded by the experience of being there, then permanently imprinted (altered or transmuted) as a result.

Check Matthew 13:33 in the Christian Bible. Jesus is quoted as saying, "The kingdom of heaven is like unto leaven." Leaven is that which causes dough to rise and grow and expand and enlarge, like yeast. "Heaven" is a Greek word that conveys the idea of expansion similar to leaven—*except* leaven denotes that which expands, while heaven symbolizes *that which has already expanded*. Even the two Bible verses in advance of the thirty-third, where Jesus speaks of heaven as being like unto a grain of mustard seed, convey the same message: rise, grow, expand, enlarge, because this is what heaven is (the realm of the expanded) and this is how you get there (expand as leaven does).

When your consciousness changes, the first noticeable difference is how expansive your thinking, feeling, sensitivity, and awareness become. Afterward, you are more than you were before. And you are transfigured and transformed to the extent that you were affected. You expand into what seems to be "heaven"—a state or realm of the already expanded (the colloidal condition).

From the power core of the cyclones (and from what could have been the inner core of a torus), I witnessed energy/matter/light being converted from one vibrational level of existence to another, recycled and renewed as if in a giant washing machine or flush tube. Whether what I saw and experienced was actually what I thought it was or not, there is more to consider here than the material we've covered.

Thanks to cosmologists like Itzhak Bentov, and to more modern consciousness researchers and technicians who employ mathematical computer simulations, this also can be said or postulated about a torus: the torus is the only self-organizing wave form. Within it is contained all the basic motions we humans recognize as movement—up/down, in/out, across. It is a living gyroscope, and the tilt of its doughnut curve, as viewed from differing angles, reflects every essential color and image design we humans have come to recognize and use plus every letter shape ancient alphabets ever came to have. This type of "information" emerges from the phases of shift the sine wave makes as it forms the torus membrane by rotating itself on its own axis.

The torus has been revealed in whole or in part in visions and dreams since time immemorial. It is the progenitor of sacred symbologies and the guardian of sacred mythologies because of the way the human subconscious automatically models death and resurrection motifs either through its shape, by function, or because of its qualities. Not only are the two cyclones I witnessed and shapes similar to them suggestive of a torus (or part of one), so, too, are the original designs of the original zodiac and the angelic and alien sky beings first witnessed and recorded by the first peoples.[31]

Fact: As a seed begins to grow, as an egg is fertilized, both are enveloped by an energy field the shape of a torus. Cut into the heart of fruit or flower and you find a torus. Speak your word and air blows out of your mouth like a torus. "In the beginning was the Word" and in all beginnings, once movement responds to its own need to move, you find a torus. Its shape allows light to stabilize and self-organize as matter.

So what's at the centerpoint? To my way of thinking, the centerpoint is a continuous state of colloidal convergence (suspension, expansion, alteration, imprinting). And we remember this state, all of us do, and we return there over and over again, because we bear the mark of its "heaven."

- In the pickle syrup story, infinite amounts of energy were at my disposal when I shifted vibrational dimensions. (My awareness slowed while my actions accelerated.)

- In the time fan story (chapter 11), infinite amounts of time and space were released once my access to the reality I resided in altered. (My awareness accelerated, while my actions slowed.)

- In the story of the two cyclones, the centerpoint to creation/consciousness was revealed once I entered its passageway. (My awareness and my actions were engulfed in a colloidal convergence.)

All three stories portray the *triune* aspects of the same state—*The Dimension of Patience*!

I became a human dynamo of activity in the pickle syrup story when I was completely and entirely at rest and became emotionally detached. Scrubbing up the mess required no effort whatsoever. I was at peace and the experience was peaceful. I knew exactly what to do without thinking about it or making any decisions, and I accomplished what I did as if in a state of utter relaxation.

Although my desire for more time was deeply embedded within me, nothing happened in the time fan story until I became completely and entirely inactive while being open and receptive. I was but a mere spectator as the world around me altered. It was not necessary for me to "do" anything when this happened except to be at peace and to accept the miracle as it unfolded.

The cyclones were part of a transcendent experience where I seemed to slip in between the cracks in my own consciousness and to enter a colloidal type of environment. This happened because of a near-death experience, after all systems and all realities had collapsed and the true me existed apart and away from my physical body.

Because of experiences that have happened to me throughout my life and from what I have learned from decades of relentless and objective research, this is how I would define the Dimension of Patience:

> Patience is a state of existence and, as such, constitutes a
> dimension unto itself. Patience, as an attribute, is a reflection

of the centerpoint of manifest reality, because the dimension of patience is the centerpoint itself. The dimension of patience is actually the "no-thing" at the heart of the colloidal condition (the seat of the collective whole in between time and space).

My reasoning follows.

When you are patient, nothing bothers you. You are either slow to react or you do not react at all. When you are patient, little or no tension is present. When you are patient, you are not threatening to others. In patience, less effort or thought is required because you automatically "know" what to do and how to do it. You act different and you look different because you are different. Patience is not just an attribute or a virtue; it is a state of existence with little or no tension, free from the restrictions that bind and bond energy to matter.

It has oft been said, "Through patience you save your soul." That's because patience helps us to relax, to release tension, to allow. We automatically become receptive to aspects of flow when this occurs, and to the signals from synchronicity that show us a connection within a greater whole has been made. Patience shifts our attention so we can meditate or contemplate or somehow realign ourselves with The Source of our being. And this realigning process can be gradual or fast, since *either* hyperalertness or the depths of deep sleep (extremes of brain wave activity) can enable us to reach "zero" and slip past the "guardians" of the centerpoint gateway.

(All mythological and mystical traditions speak of some type of guardian whose job it is to block an aspirant from reaching higher states of consciousness or "heaven" before that soul is ready. *The "guardians" that block further entry into transcendent states are simply our own perceptual preferences.* We block ourselves from reaching our goal by what we think and by what we believe to be true.)

To slip in between the flashes of on-off reality, in between the oscillations of energy's need to vibrate, past all subjective and objective levels of recognition and awareness to the realm where no-thing resides, we still ourselves and enter the Silence (which leads to the Light of All Understanding).

How do you solve a problem? By forgetting about it and doing something else, resting, then the answer pops in. Troubled? Lose yourself in service to others and your problems solve themselves. Need inspiration? Blank out and enter a flow state.

Directions change when you discover what lies in the Silence. Directions change because you change the vibrational speed of your own brain when you rest or when you lose yourself to such a degree that it seems as though you have rested.

When you change speed, so does your world. Change speed fast enough, suddenly, even violently, and you could pass through the Silence and enter a colloidal state that expands into the universal collective. The experience leaves its mark. You know more because *you suddenly know more*, and you know that you know and you're never the same again.

So how do you recognize when you are slipping past the cracks of consciousness into the In-Between? When tension disappears.

It's tension you see, tension to whatever degree that keeps us held in check, that keeps us embodied, structured, confined to the limits of our own perceptions. Without tension, any of it, nothing can prevent us from moving past all restrictions and expanding back into the All. What do most people say once they have undergone a dramatic transformation that resulted in full or partial enlightenment? "I felt as if I had found my true home, where I really belong, perfect bliss." And what do most of these people exhibit in such abundance afterward? Profoundly more patience.

Tension is what keeps light solidified as matter and upholds existence and makes creation possible. Without tension, form would have no structure, no integrity, no specific shape, no potential for variation; there would be no "fabric" to the universe.

Creation's beginning may not have been so much a big bang or a big blast as the bursting forth of a great thought that stirred; and as that thought moved, time and space spun off and the torus rolled forth and there was tension, the tension of thinking the great thought into place, the tension of the created responding to The Creator, not once but continuously.

Cayce said there were three dimensions: time, space, and patience.
Cayce was right.

Patience is the colloidal state of the In-Between, that third dimension of existence that allows everyone and everything to converge and unify, to remember and know. It brings together that which time and space separate.

Patience frees consciousness to awaken to its source and, in so doing, uncovers the web.

14

Secrets in the Web

*The great secret in life is in learning to see things in their wholeness, and
to realize the inside and the outside simultaneously.*
—Thomas Troward

Beneath the visible world of mute matter and seeming separateness,
there is a deeper web of reality. This deeper place cradles not only
the phenomenon of future memory but the common denominator that
enables consciousness and creation to coexist and cocreate. That common denominator is the colloidal condition.

I have found that meditation, sudden shock, trauma, paranormal
episodes, near-death experiences, religious conversions, spiritual transformations—all have the same potential to trigger a colloidal state. We
recognize that such a state not only interrupts and then alters consciousness and the perception consciousness registers, but it also frees unnecessary tension so human potential can expand.

*This expansion enables more information to exist, which increases
intelligence.*

In other words, whenever the potentiality of consciousness expands,
we retain the imprint of that expansion, and the imprint we retain is the
information we gain, *the knowing.*

Question: Where does the knowing come from?

Answer: From the same source that tells a cell what to be.

Microbiology has demonstrated that DNA molecules tell cells what to do, not what to be. Yet all cells know their identity and their purpose before a single task is initiated. They display advance knowing. Particles and light waves consistently register knowing in advance, as well. For instance, a particle can jump into another orbit without passing through the space of that orbit, and know exactly where it is and what it is to do without the experience of having traveled to get there (this is referred to scientifically as a "quantum leap"). The earth and its myriad life-forms behave and operate in the same manner, as if connected to a single source of guidance, a unified intelligent whole.

Information, then, preexists in some form. This preexistent, basic information seems everywhere present and readily available; waiting to be tapped, to be accessed, to be used.

Such pure, changeless, yet ever-changing information may well exist in invisible etheric fields, making up something akin to a collective "memory bank" or etheric "library." Entrance to these fields would most likely be spontaneous, either from receptivity to the field or by intention, or because of need. (Remember my experience in 1978 with the Shenandoah River and my encounter with fields upon fields of memory "libraries.") Freud labeled such a memory bank "racial memory"; Jung called it "the collective unconscious." Yet, maybe it is something else, something more.

Check out these facts:

- Back in the thirties and forties, Harold Saxton Burr, Ph.D., professor emeritus of neural anatomy at Yale University School of Medicine, established that *after* conception energy expands around salamander eggs nine hundred times greater than normal, thus marshaling available energy from the universe at large to create a massive energy shield for the protection of new life formation. He discovered that this massive energy contained within it all the information necessary to direct cellular creation, literally a blueprint. Burr was heralded for his discovery until he stated that the same thing happens *after human conception with the human egg.* Suddenly all grant monies were canceled and he was left to face criticism so severe, he never fully recovered.

His books, *Fields of Life* and *Blueprint for Immortality*, now classics, are as relevant and timely today as when first written decades ago.[32]

- In the early eighties, Rupert Sheldrake, an English plant biologist, used the term "morphogenetic fields" or "M-fields" in his provocative book, *A New Science of Life*,[33] to describe a phenomenon similar to what Burr discovered. Sheldrake posited that M-fields direct shape, development, and basic behavior of living species and systems, functioning as invisible blueprints, or connecting memories, linking any member of a given species or system with its fellows, while supplying a ready bank of accumulated information. He claims these fields constantly change and evolve as individual members learn and grow. Changes in the field are not always instantaneous, since, according to Sheldrake, a certain threshold must first be reached before the entire field responds. Once the total number of changed members reaches that threshold, he believes field activation occurs and all members, no matter where located or how much time has lapsed, are then stimulated to copy or somehow express the change. Sheldrake's theory hasn't fared much better than Burr's in the scientific community. At this writing, however, all attempts to prove it false have failed, adding, instead, to the mountains of data supporting it.

Indicated in the work of these scientists is the idea that a cell knows what to be because the pattern of its shape and the purpose of its being already exist on a level of etheric "blueprints" (cosmic memory banks). This suggests that once the energy charge of conception "sparks" cell division, the soon-to-be-cell locks in by frequency of vibration to whatever pattern it resonates with. Cohesion results, along with the instantaneous attraction of an energy-rich membrane or plasma shield of information that nourishes and directs DNA growth.

Question: How then does this information (knowing) travel?

Answer: Through a fabric netting or web, a web that connects the universe to the information needed for its continuance.

To set the stage so we can probe this idea further, know that near-death survivors and others so transformed often comment on microscopic "threads" of luminous, vibrating light they see connecting everything and everyone together in a giant "web." Many of these people go on to claim that such threads explain the interweavings in life and the interconnectedness of all creation.

They also say that once light pulsates harmoniously through any section of the webbing, regardless of distance or time span, synchronicity naturally results, signaling that one portion of the network now resonates with others of equal light.

Along this same vein, there is a teaching in the Huna philosophy of Hawaii and Polynesia about invisible threads capable of connecting people, places, and things. Referred to as Aka Threads, these flimsy filaments (threads) of vital force or vital life substance (Aka) supposedly extend from and beyond our physical body if we touch anything or move around. Like cosmic fingerprints or etheric rubber bands, these streamers spread out in threads so light and airy they can easily be broken or dissolved unless continuous touching and movement, or frequenting the same environment, reinforces them. The more often the same thing is touched, the thicker the filaments get. Subtle in effect but supposedly durable once thickened into cords, Aka Threads are said to link or bind things together, even to influence soul relationships from lifetime to lifetime.

Other traditions from various cultures across the globe have similar beliefs, and each describes the same basic type of invisible, interconnecting threads, or energy webbing, created by life-forms interacting with each other. Such traditions further claim that these mundane, interpersonal threads serve to reinforce the overall threading of the entire fabric net existent throughout the universe since the beginnings of time and space.

It has been my experience that, although completely elastic, this normally imperceptible latticework is deceptively strong and powerful. It must be, for the fabric constitutes the matrix that time and space collapse into when energy expands.

When internal to us, I would define this matrix as "consciousness."

When external to us, I would define this matrix as "ether" (also spelled "aether").

As I see it, external variations of the matrix (ether) are just as intelligent and knowing as internal versions (consciousness). This weblike matrix, external to us, emanates from the centerpoint (the colloidal "hole in the doughnut") and carries its imprint (the expansive potential of unfolding information).

Yes, I am aware of the fact that no recognized scientific experiment has shown any trace or suggestion of an ether. Yet for centuries, learned ones have argued its existence. Look up the word in a dictionary and a gas or a liquid is all that will be described plus fleeting reference made to a hypothetical notion that this "substance" was once thought to occupy all space. An English physicist, though, is about to set the record straight.

Harold Aspden, a professor of physics and former head of IBM's patent office in Europe, has written *Modern Aether Science*,[34] a book many now believe should be required reading as an introduction to college physics. He uses standard electromagnetics to explain the phenomenon of ether and says that its effects have always been present but either went unnoticed or were ignored. "There was impatience at the difficulties of fathoming and charting that sea of energy permeating space—the aether," Aspden states. "And so, many pretended that the aether does not exist and did so by abstract mathematical formulations."

Thanks to Aspden, ether is once again a legitimate topic for discussion and one I intend to pursue with vigor. That's because I'm convinced ether is real and that I have seen it.

And what I have seen I would describe as a pliable, closely woven, preatomic web of interlacing, interacting, circular logic units (airy threads of bubblelike components), which make up the fabric of information (intelligence) that allows thought and the "flow" or "movement" of thought to expand, contract, stabilize. These circular logic units compose the ether, and they can appear as sparkling or twinkling particles or sparks, or somehow inert and dark (depending on whether they're "on" or "off," visible or invisible). The ether I have seen can vary in contour: thinner near objects of mass, and thicker farther out in space, where its denseness constitutes what I believe is zero-point energy.

We learned about zero-point energy in high school. In case you have forgotten, here is a refresher:

Each cubic centimeter of space, including that of a perfect vacuum, contains a stupendous amount of untapped electromagnetic energy. This energy is conservatively estimated to be on the order of nuclear energy or greater—primordial and unlimited. The term "zero-point" is used to describe this energy (since at absolute zero gas has no volume). It is known that the universe in which we live measures three degrees above absolute zero.

As a brief aside, I was privileged to visit the Atomic Energy Testing Station near Arco, Idaho, four times during the late fifties and early sixties. Three of those times I participated in general public tours during annual open-house programs held there, and the fourth was with the Air Force Association's membership tour of the hanger where the atomic plane was to have been built. Although the atomic plane project was scrapped as impractical, I use the word "privileged" to convey how touring the nuclear facility affected me. I loved the place.

Not as much was known then about atomic energy and fission as is known now; hence, visitors such as myself had quite a bit of latitude. Yes, we were required to wear radiation badges while on tour, and yes, we were checked regularly by having to walk through doorlike detector frames. Yet, reactors were only off-line (shut down) for a few hours before the crowds came and always there was one, namely me, who took advantage of that. Since part of each tour included the opportunity to climb stairs to the top of a still-somewhat-active nuclear reactor with its cap opened and peer inside, people like myself, if clever, could easily outwit on-duty technicians. And this I did, consistently, for three years running.

I couldn't get enough of it. When my turn would come to peer inside the reactor core, I'd stick my whole head in there and stay as long as possible. I'd drink it up, smell it, taste it, bliss out on the energy. I could even see it, the vibration of the core's radiation, and to me it was a fresh, sparkly bright, piercingly wondrous radiance so utterly pure and powerful it was ecstasy itself even to be near the stuff. As I walked around the reactors, I would caress them, kiss them. Even when I stood over pools of irradiated matter, I could still feel the vibrations and I could see, smell, and taste them as well. I would devise ways of avoiding detector checks so I could go back and absorb more of that energy and absorb I did, like a sponge. I called it LOVE, that feeling the energy gave me, LOVE in capital letters because it was a feeling beyond any frame of reference I had at the time. When public tours of the facility ceased, supposedly for "budgetary" reasons, I was crushed. This cancellation meant to me that I would forever be denied the most incredible source of love I had ever known.

A comparison can be made here. Of all the things I have ever encountered on the earthplane, of all the feelings, passions, and sensations I have ever felt in or out of the body I wear, the one that is most like that feeling of love I encountered on the other side of death is the feeling of that brilliant radiance I was repeatedly showered with when I stuck my head in an atomic energy reactor off-line but a few hours.

I do not mean to imply here that nuclear energy or its radiation is harmless just because I suffered no ill effects, but I do mean to imply that the ecstasy and the unlimited freedom I experienced from the love I encountered during my near-death episodes is strikingly similar, if not the same, as the exposure I received from the fission process. (This was especially true during my third near-death episode when I was confronted by the two cyclones and drawn toward their middle—the centerpoint.)

To put it another way, the radiance I encountered at the centerpoint (a subjective reality during an internal, transformational experience) and the radiance from those reactors (an objective reality during an external, physical experience) match that of the ether matrix (a primordial issuance from the collective whole).

The radiance is the same, because for all intents and purposes, the power source is the same.

Ether to me is weightless, transparent, frictionless, and permeates all time/space/matter. And I believe, as so many others do, that it is the medium by which light can be transmitted in a vacuum. The microwave background of its circular logic units is even throughout, suggestive of order, purpose. The ether matrix has tides and flows as if its fabric threading were "floating" in a type of plasma, a primordial "soup." The power of zero-point energy seems to be its major "ingredient," along with dark (invisible) and luminous (visible) matter.

Einstein, too, was convinced that the universe must exist within a type of fabric matrix. In his concept of space, celestial bodies behave as if they are on a "rubber sheet," with the force of gravity similar in effect to when a heavy object is placed upon the sheet, bending it. Any lighter object will roll across the sheet toward any heavy object. The denser an object, the more it bends the sheet. In other words, the rubber sheet, the fabric, is completely flexible.

(Einstein's concept highlights the fact that the ether fabric thins out when near objects of mass. This thinning supposedly allows gravity [the entrainment of attracting forces in motion] an opportunity to respond similarly to how a bloodstream protects and nourishes its host. It seems to me that the wave effects created by gravity operate more like containment shields for plasma than like any type of force. Quite possibly, gravity's task is simply to hold together via entrainment that which has the best potential for further development. To better understand what I'm suggesting, study hydrodynamics and the characteristics of compressible fluid. A question to ask is: Why is it that the fact of a comet's tail always being pointed away from the sun is used as evidence that there is a solar wind, when the direction of a comet's tail may actually be an example of a plasma wake—not that of a solar wind?)

What really convinced me of this fabric, the ether matrix of the universe, was a lifetime of snatches and glimpses of its reality. At the age of four, for instance, I could clearly see the ether, its tidal forces and wave forms, in the sky and in the air around me. Over time, I learned from studying the flux of those airy patterns how to accurately predict atmosphere and climate (external weather) and people's moods and general health (internal weather) in any given area where I was. Sometimes in play, I would cuff ether's gauzy threads just to see what might happen to the fabric when I did. This ability of mine to see and interact with what no one else saw created quite a stir in my neighborhood and among my peers.

As an adult, I lost this ability and even forgot all about it until the pickle syrup incident inspired me to rethink the nature of existence. The events surrounding 1977, however, required that I embrace a much greater vision of what life may be in order that I might understand my near-death experiences and the aftereffects that followed.

Remember, when I stepped forward to stop the overflowing pickle syrup, I came in contact with air the look and feel of threaded fabric, reminiscent of what I had experienced in my youth. When I described the mistlike fountain forming on my drafting table, I could just have easily used the term "gaseous" instead of "mist," for it was like a weightless, buoyant plasma. When I saw the cyclones, I also recognized that everything around me, the light-filled world I had entered, was filled with an interactive matrix that "floated" what I saw, *while at the time participated in and with what I saw.*

And of all the many times I have been in The Void, always, there has been and still is a radiant presence of shimmer and an awareness of conscious intelligence as if The Void were alive and breathing.

Because of my broad variety of exposure and experience, I have come to regard ether as

- Ectoplasm (when physically manifested): That gaseous-type substance said to exude in gauzy streamers or buoyant mistlike clouds from a medium's body openings (mouth, nose, ears) when the individual is in deep trance and channeling "proof" of life's continuance from The Other Side. During such sessions (or seances), the appearance of ectoplasm is considered a sign that the substance of the universe is passing into and through the medium for the purpose of revealing the validity of alternate realities. Totally flexible and capable of taking on any shape, the substance can, according to tradition, compact into a solid and remain solid as long as the power enabling it to manifest lasts. Many phantom apparitions and strange lights consist of ectoplasm and, in all probability, ghosts.

- Pure Spirit (when not physically manifested): That presence of indivisible consciousness everywhere existent, known biblically as the Holy Spirit, metaphysically as "the numinous." The root word of "spirit" means "breath"; esoterically, "spirit in matter" refers to "the breath of life breathing." The power of pure spirit (very probably zero-point energy) is called *qi* or *chi* by the Chinese, *ki* by the Japanese, *prana* in the yogic traditions of India and Tibet, *mana* throughout Polynesia and Hawaii, *baraka* to the Sufis, *orenda* to the Iroquois, *megbe* to the Ituri pygmies, and *yesod* in the Jewish cabalistic mysticism (to mention but a few of the over four hundred different names used to describe this; names vary according to culture). The late Wilhelm Reich termed it "orgone."

As near as I can tell, the field array of "memory" (circular logic units or preexistent information) in the ether fabric is aware of its own reservoir of intelligence, hence linking the created with The Creator. But information does not travel, even in the ether. It just seems to because of the wave effect of gravity and plasma, and the "dance" of light's varied reflections throughout manifestation as each projection glances off the curves of countless tori. (Cartoon movies operate on a similar principle, for in order for the cartoon characters to appear as if they are moving, thousands of single-frame drawings must "flip" by a light source at high speed.)

Information preexists and is everywhere and equally present (that is, the equal distribution of microwaves throughout the matrix). Since information never goes anywhere, you access it or it accesses you through the web's threaded networks. Interchange, such as a deletion or an addition to its reservoir of memory, cannot happen without some type of stimulation.

This means that

- evolution describes the increase and the refinement of accumulated intelligence in any given field of memory storage.

- devolution refers to the decrease or stepping down of such intelligence.

Because of the construct that supports and maintains the illusion of time and space, energy is seen as moving, not information. And the movement of energy seems in continuous waves ("analogue," connected together), when actually what exists are particle sparks akin to digital bits ("discrete," separate, and seemingly detached units).

Of fascination is the scientific fact that the "flow" of these digital bits is not smooth but comes in interrupted packets called quanta. Physicists tell us that the interval between transmissions of these energy packets is based on a ratio of six to one, six flashes for each rest. The Bible tells us that for six days God labored to create the universe and on the seventh day God rested. However inexact biblical language, the formula is correct. After every six transmissions (activity), there is a pause (rest).

(Visionaries, like those credited with writing various sections of the Bible, regularly pass through the In-Between and witness or access the true reality behind creation. Although they cloak their revelations in a language steeped with symbolic imagery, the message they seek to convey is often quite accurate.)

Truly, it is our perception that sees energy as flowing in waves when what actually exists are logic units blinking on and off a zillion times per second between what appears as existence and nonexistence. This blinking allows light and darkness to create contrast, a necessary element if we are to have a meaningful experience in the earthplane. Without contrast, interactions cannot be recognized nor can interrelationships take place.

What I have noticed on a mundane level, however, is that energy originates in what appears as "activity" from the way it stimulates the memory of preexistent information. This memory exists within the fabric's webbing. To put it succinctly, jiggle the ether and things happen. (Memory's imprint activates the webbing once it responds to the jiggle.)

It's like magic the way this can work.

Example: I never told you how I got my job in Washington, D.C., while "acting out" my first encounter with the future memory phenomenon:

> I was down to my last ten dollars and one tank of gas. No prospects. No more money. No visible alternatives. I had visited with David McKnight that weekend in the Shenandoah Valley to celebrate the festival of Oktoberfest (he paid for my gasoline), and had returned to Washington, D.C., Sunday night. After meditation Monday morning, I suddenly knew I would find my right employment and start work on Tuesday. That knowing was almost euphoric in quality, and it seemed to guide me directly to a specific employment agency.
>
> After passing the agency's tests, I was sent to interview with a firm located on the third floor of an office building on

the corner of Twentieth and I Streets. The moment I walked through the open door into the firm's suite of offices I saw myself already there. The "me" I saw was active and real: sitting at the desk, answering the phone, walking in and out of various offices, going back and forth to filing cabinets, typing.

I was so surprised by this scene that all I could do for a while was stand at the doorway and watch "myself" busily at work. But the me I saw was phantomlike, clearly me and fully dimensional, yet of a sparkly gaseous substance. The me I saw was composed of manifested ether (ectoplasm), and that etheric form knew exactly what it was doing. I had to walk right through my phantom self to reach the room where I was to be interviewed. As I passed through it, my phantom self felt similar to a cool, gauzelike mist, and its odor was that of a faint, yet flat, ozone-with-a-tinge-of-ammonia smell.

Suffice it to say, I applied nowhere else. I reasoned that, since I was already working for this firm, I might as well make my employment official.

My need for a job attracted more "energy" to me in the way my emotions responded to my need. But my choice to take this need of mine ("energy") into meditation and turn the situation over to God enabled the emotional energy I had produced to become the jiggle necessary to access the ether fabric of information. The result was a partial colloidal state wherein my previous reality collapsed and a new reality emerged. I was euphoric, which is similar to the thrill one feels when suspended between time and space, and I was possessed of knowing because information had been accessed. In this case, though, the information that was accessed physically manifested as ectoplasm—ectoplasm that formed around and took on the "spirit" of the preexistent information that reflected my future. My walking through that particular doorway initiated the projection of that ectoplasm. Because I was there to witness this feat, I interpreted what I saw to be a sign that this job was indeed the right one for me, so I accepted the job offer. And the job I accepted *duplicated exactly* what was previously reflected in ectoplasm.

This episode became the first of many such occurrences I have had whereby information's imprint (whatever is remembered, learned, or accessed from the greater reservoir of intelligence) manifests itself in response to being accessed. Although this incident could be called another type of future memory (as it involved what seemed to be my future), I would differentiate it from future memory episodes previously discussed in this book.

The differences are as follows: Dimensions did not alter in any manner (I was not presented with nor was I involved in a futuristic preview of this world while resident in another realm); activity was ongoing (my physical body was not immobile or "frozen" but fully animated); my consciousness had not altered (I was still engaged in the same train of thought); the phantom me was industriously doing its own thing (I was not connected to it, nor were its feelings and awarenesses mine); plus there was a clear distinction between the two of us even though we were both physically manifested at the same time, in the same place, and in the same dimension. (The ectoplasm version of me looked more akin to a colorless, gaslike hologram devoid of emotion, while the solid-appearing me displayed the colorful and complex arrangement of molecules and atoms enjoying a rush of excitement.)

My sense of what happened in this incident is that I somehow "bumped into" or discovered what many people call "fate." To put it another way, I bumped into *my blueprint self engaged in an activity in advance of what my manifested self would later carry out.*

If the Australian aborigines are right in their understanding of existence, and I think they are, then the two parallel space/time continua they identify as dream time and wake time must have merged for me that morning—enabling my blueprint to commingle with its manifestation.

Impossible?

Just wait till you turn the page.

15

The Thought That Stirred

These sparks, human souls, which come directly from God, have no end:
they are imprinted forever with the stamp of God's beauty.
—Dante Alighieri

D avid Bohm, a professor of physics at Birkbeck College in England, declared that the incredible interweavings throughout manifestation demonstrate "the implicate order" (folded together and intertwined, like unto a cosmic sea of energy), where any relatively independent element contains within it the sum of all elements (the part reflects the whole).

Bohm's concept of the implicate order is based on twenty years of test results from both mathematical theories and laboratory experiments. These tests showed how subatomic particles are able to respond and relate to one another in ways not explainable by the law of cause and effect. He is quoted as saying, "Thus, one is led to a new notion of Unbroken Wholeness which denies the classical idea of the analyzability of the world into separately and independently existent parts." Bohm found that on the subatomic level Newtonian fundamentals are no longer completely reliable. At this level, all things are interrelated, intertwined, and intimately connected, because physical reality is threaded together in a seamless web of responsive, conscious intelligence.

This means that

- Interrelationships can defy the law of cause and effect.

- People and particles and all aspects of life can influence each other without direct provocation.

- Once anything has interacted with another on any level, both parties can then continue to respond to each other without further interaction.

- Time and space make no difference, because in a way invisible to human perception, at the most basic level of existence, everything is connected to everything else and the scientific laws of probability do not always apply.

The Einsteinian Revolution established that (1) relationships are more real than identities; (2) nothing is separate from anything else; (3) all things flow into each other; and (4) past, present, and future have entirely different meanings than we might suppose.

Add the three dimensions science recognizes (height/depth/width) to the fourth described by quantum physics (relativity), and you get what Edgar Cayce spoke of when he said three dimensions governed earth—time (height), space (depth and width), and patience (relativity). Through the dimension of patience and the receptivity patience fosters, the wonders of spirit are unveiled, the "imprisoned splendor" in and beyond the centerpoint, a splendor forever seeking release in a desire to experience and express itself.

Truly, infinity is not bound by the concepts of those who seek to understand it, for *the only reliable geography is consciousness*. And it is consciousness that composes the "space" that cradles light and gives it being.

Speaking of consciousness, a new theory about the subject has taken the scientific community by storm and it deserves our attention, especially since I named consciousness as the internal aspect of the ether matrix.

This theory centers around "tachyons," a term taken from the word "tachus," which means "speed." Tachyons are hypothetical particles that supposedly move at speeds faster than light, a feat possible because it is said that they possess no mass and no electromagnetic fields. Their spin rate is classified as *in between* positive and negative charges, thus giving them the ability to produce order out of chaos (called anti-entropy by science). No one has yet demonstrated the existence of these speedy and unusual particles, but Jacques Steyart, a physicist from Louvain University in Belgium, has already observed electrical effects from them.

Tachyons are thought to exist within an oscillating, energy-rich field, or neutrino "sea," where everything "floats." Neutrinos, by the way, are elementary particles without electrical charge and too close to absolute zero to be measured. It is believed that tachyon fields lie beyond the light barrier and are superluminous (possessing an unusual brightness without known cause).

Tachyonic superluminous time-space (the tachyon fields) differs from regular time-space. Time and space merge at the vibrational frequency of the tachyon fields, supposedly allowing instant access to existent and pre-existent information, while fostering a simultaneous-everywhere type of consciousness and an instantaneous type of telepathic communication, *for consciousness itself is said to be made up of tachyons.*

The theory further postulates that thoughts can overlay and overlap each other in a field of tachyons. And this exactly illustrates what Itzhak Bentov was referring to in chapter 9 with his convergence chart of objective and subjective time-space states, where "we can fill the entire universe with our consciousness at infinite speed and be everywhere at once."

Since these superluminous particles seem to *increase order and information* when disorder or decay breaks down into entropy (inertness), the moment of physical death (especially with human beings) could well be like a transition from *subluminous* consciousness (regular space-time involvement), through the limbic light gateway, to *superluminous* consciousness (reunion with the greater whole in and through the tachyon fields).

Doesn't this remind you of the typical description given by the average near-death survivor about what he or she experienced in dying?

If indeed consciousness is made up of tachyonic particles and this can be demonstrated, then the doorway to spirit worlds will finally be given the recognition it deserves.

Consciousness and ether, mirror reflections of the web's matrix, are in essence the same to my way of thinking. Whether existent as tachyonic particles or as blinking circular logic units, and whether connecting or overlapping subjective and objective realities, the same fact emerges: matter and energy are manifestations of information, and so is light and so is consciousness and so is ether. As available energy is used up (disorder and entropy), information increases and the intelligence behind and undergirding information expands. Nothing is ever lost, because everything is recycled and restructured. It's as if only one mind experiences and evolves, learns and grows, then goes "home" to itself.

Looking at the universe, what we can see of it, may really be like looking at *the inside of a giant brain processing thought*—the plasma of ether its nourishing fluid; gravity its blood vessels; the threaded web and the superluminous tachyonic particles its neural network and synapses; the objects of dense matter (such as solar systems, planets, dark matter, and life-forms plus the lives they lead) the manifestation of its thought-forms and thought-form potentiality; the stirring of its thought the oscillating pulsebeat that creates time and space. (This dovetails what I witnessed during my third near-death episode.)

The symbiotic relationship of the whole to its many parts becomes clarified when we consider that The One may need all of Itself to experience Itself—that in actuality there are no graduations in life, only the celebration of new beginnings.

One mind, many thinkers.

Information is preexistent because only one mind exists, one source of all things, one power behind and beyond all processes, one force in and through all as all, one intelligence.

God.

And what do people who have undergone a near-death event or a spiritual transformation or any type of enlightenment or conversion experience talk about first and foremost?

God.

Their number one discovery is that God exists. Their number two discovery is that we are indeed made in the image and likeness of God in the sense that, as souls, we are sons and daughters of The Most High, children of the same God, extensions from the same Source, cells of the same Body, unified and one because we are of The One—with the same creative potential in consciousness as our Creator.

Most near-death survivors and others so transformed grow younger after their experience and become happier, and why not? To have glimpsed a grander vision, no matter to what extent, is a wondrous thing. To realize we are beloved thoughts in The Mind of God is the *real* heaven.

As Hermes the ancient Egyptian priest once wrote: "Then, in this way know God; as having all things in Himself as thoughts, the whole 'Cosmos' itself."

Because this truth is *so* important, and because the exploration that we are engaged in of future memory and the innerworkings of creation and consciousness is so provocative, I want to share with you the panorama of creation's story as revealed to me during my experiences with death. Think of it as you will.

The Thought Which Stirred

Beginnings

Creation is the stirring of thought in The Mind of God. Colloidal in function, that stirring creates the energy pulse that sets in motion and maintains the ongoing oscillation of time and space, and the ongoing unfoldment of the thought which stirred. The makeup of this thought is consciousness and all that results from its stirring bears the mark of its consciousness. Creation allows The One to become the many.

Spirit Creation

The suspension of the colloidal state establishes the plasma matrix, a subtle field array or sea of unlimited energy and power. Seemingly mistlike or gaseous in form, the plasma matrix is actually a conglomerate of shimmering bubbles. The plasma matrix (spirit) is the passive aspect of the thought which stirred. It propagates itself easily and effortlessly into the fabric of creation that issues forth throughout all resultant levels, dimensions, and realms. Since it is possessed of knowing, this etheric spirit matrix (ether) is the receptacle of information from the colloidal state.

Soul Creation

The expansion of the colloidal state releases the active aspect of the thought which stirred in the form of charged nuclei of sheer power or sparks. Each has volition and the same potential for intelligence and the ability to use that intelligence as the thought which stirred. Each is created in the image and likeness of its Creator in the sense that the Creator originates all intelligence, all mind, all power. Hence, the nuclei possess the imprint of divine potential. Although seeming separate and singular and issuing forth as showers of sparks, the nuclei actually spread out and expand in waves, thought waves, similar in movement to what happens to the wake caused by a speedboat zooming across a placid lake.

Soul Formation

As the thought-wave projections move ever outward and farther afield in the matrix of etheric plasma, each nuclei inside each wave divides into positive and negative poles of charge, mimicking what causes the oscillation of energy, and then spins off. The nucleus, once divided and freed from its wave, can take on any manner or type of existence or shape. It increases in density the farther removed it becomes from the essence of its origination. Yet, no matter the experiences it takes on, the nucleus spark carries within it the urge to attract unto itself that which will unify its charge (the rejoining of its positive and negative poles back into the full nuclei it once was), and fulfill its ultimate purpose (reunion with its original source). The many ever seek to rejoin with The One.

Soul Projections

There is no limit to how each nucleus spark might project itself, combine with others, grow, or alter. Each is powerfully charged to seek experiences, as experience leads to information and information expands with experience. Those who allow themselves to become large and dense participate in the formation of celestial bodies, while those who desire more flexibility experiment with different ways of shaping life-forms. Once each nucleus expands and enlarges its information base, it can and often does reproduce itself by projecting extensions. Seldom will any soul nucleus lend all of its power to a single extension, since to do so might overload that extension and destroy its form (thus eliminating the very experience that inspired the extension's projection). Any reproduction from these souls will carry the same imprint as the nucleus has; the same urge to experience and grow, then reunify its forces and rejoin with The One True Source.

Embodiments

Soul extensions are embodiments. These usually involve group efforts and evolutionary-type arrangements. Form, shape, manner, or purpose of embodiment are without limitation, as the soul's "birthright" of divine power fills each. Responsive personalities are often created so embodiments will link more closely with the soul's energy, though direct contact with The One True Source is equally available to each and all (for embodiments extend the imprinting of consciousness). Those souls who interest themselves in the variations possible through the development of responsive personalities most generally commit to humanoid forms. Although different variations of humanoid embodiments can involve emotion, the human pattern developed in the earthplane offers a wide range and rich diversity of emotional sensations. This is why so many souls seek human or humanoid embodiments on earth or whatever is similar.

Reconnections

Seldom do embodiments of a soul's spark enable that soul nucleus to rejoin with its other half, but when that does happen, tremendous amounts of energy are released and the original charge of the nuclei reunifies. This reconnection of "twin" souls usually happens on levels of experience other than embodiment. However, embodiments can and do reconnect the soul nucleus with other members of its original thought wave fairly often. These joinings are companionable and sometimes pleasurable because of shared memory, and they foster "soul mate" relationships. A soul can have numerous soul mates but only one twin soul. Thus, the urge to reconnect is actually threefold: with other members of the original thought wave, with the other half of the nuclei power charge, and with The One True Source (The Mind of God). Reconnection is love fulfilled, for love is the recognition of The One within the many; it is that force that bonds and unifies and makes creation possible (not an emotion per se, although it can thrill as if it were an emotion). The power of love enables the lesser self to rediscover the Greater Self, the core power behind and beyond the thought which stirred. Literally, love is the reality and the expression of God's Presence.

Cycles

Embodiments are purposeful and usually planned. Reasons for them are specific and follow the patterns of growth and learning possible in and through experience. Differing embodiments from the same soul nucleus can happen singly, as parallel experiences, in multiples or combinations or simultaneous. Limits do not exist as to how embodiments can be arranged or carried out. This is also true with human and humanoid life-forms. Reincarnation in the sense of progressive or successive lives is misleading, as there is no such limitation on soul embodiments. There are experience cycles, however. These cycles evolve around an overall theme and are planned according to whatever is necessary to carry out and fulfill the theme. For instance, if the soul wanted to explore the nature of courage, various experiences and different life opportunities would be planned that would enable its embodiment(s) to embark upon that type of exploration. It would not matter how long the exploration of courage took, since time has a different meaning to a soul than it does to an embodiment.

Fate and Free Will

Thematic cycles can be fulfilled in one embodiment one time, can involve a series of lives for the same embodiment, or can be worked out through various types of multiple embodiments and differing life opportunities. Usually a given embodiment will have successive lives in the sense that its basic personality construct will evolve through the cycles, building on experience as it goes along. To this degree, reincarnation (life-after-life progression) may seem "the way of things" when that is not necessarily true. "Fate" is a term used to describe those specific and planned events encountered during specific and planned cycles. These seem "fixed" in the cycles because of choices made previously, and they usually remain as planned until met (fulfilled) or completed to whatever extent. But that which seems fixed still carries within it an open range of responsive choices. Variations of response can alter events or change entire scenarios. This is because choice is inviolate. As the soul has free will, so too does any embodiment. And with that freedom of choice, the embodiment can choose to cooperate with or deny its soul, oppose or deviate from any plan its soul has made. Life experiences tend to run smoother or at least be more effective and satisfying if the life-form cooperates with its soul and accepts the plan of the cycle. The choice factor, however, is layered with possibilities and is not limited to any particular result or concept of sequential timing.

Enlightenment

The need to separate and the urge to reconnect compose the one great paradox. There are the many, yet there is only The One. With life-forms, sex is an expression of that paradox. The orgasm it offers mimics the colloidal state of creation. The instinct for sex is just as much a part of the imprinting as any procreative or creative urge, since all such urges are essentially the same and spring from the same longing. Certain drugs can mimic the memory and the power of the colloidal state as well. Both sex and drugs are short-term solutions to the paradox; neither satisfies nor solves it. The bliss of enlightenment (the ultimate orgasm) is different. This transformational shift in consciousness can so alter and change a life-form that long-term satisfaction and solutions to the paradox can result. Although enlightenment is multileveled in the way it affects both an embodiment and its soul, the real goal of enlightenment is neither religious nor spiritual in nature; rather, it is to unveil the original imprinting made on the nuclei from the thought which stirred. Once this occurs and Truth is recognized, accepted, then integrated, the embodiment is freed from all former and present entanglements as it frees itself to either rejoin with the full power charge of its soul or expand into a soul nucleus of its own making. The latter is true reproduction.

More Beginnings

Creation is ongoing because it is enjoyable. All embodiments are gods in the making because each and all can free themselves from the projection process and rejoin with or independently expand into a soul nucleus. All souls are gods in the making because, once transformed by the experience of thinking of themselves as separate, they can reconnect with their greater power charge and reunify in a thought wave (either their original one or another). Fully self-realized and reenergized thought waves become companions with The One True Source in the sense that they are now ready and able to develop entire universes from the spirals they emanate from the original spiral of God's Thought. *These waves are the thought which stirred. They are the paradox solved. They are The Mind of God within itself as itself expressing itself.*

I have seen them, the thought waves, and they are beautiful beyond beauty, lustrous, shimmering, gloriously luminous and radiant, musical. And, oh, the tones of their music. To hear the sounds of their tones is to forsake forever all else save the memory of God. (Hoomi singing comes the closest on the earthplane to this type of music that I have yet discovered. Refer to the resources section for information about it. Also of interest is the child's book *Children of Light*, which depicts the holy sparks of Light coming to earth to experience the power of choice, then returning to their Source much wiser than they were before.)[35]

In a strange kind of way, J. R. R. Tolkien wrote of this celestial music and of the thought waves, only he couched his story in mythic imagery for his book *The Silmarillion*.[36] Find a copy and read the first four pages of the chapter entitled "The Music of the Ainur." There is no need to quote his text here except to say that Tolkien used the idea of musical tone waves, as a representation of the stirring of God's thought, to explain how creation poured forth, and how, from the themes of God's thought harmonics, the created "Holy Ones" (souls) produced and manifested their own diversions—melody an expression of "good"; dissonance, "evil."

As Tolkien's book describes the beginnings of creation and the separation of the holy souls from The Mind of God, the children's movie *The Dark Crystal* illustrates the reconnection possible when the holy souls reunite with their original charge (before the nuclei separated) and return to their Source. Rent a video if you can and study the story line.

Yes, this marvel of puppet wizardry runs through typical themes of good and evil, sacrifice and greed, but it also emphasizes that *neither* duality is valid, that appearances can deceive. In the closing frames of the movie, the evil monsters and the blessed wise ones merge into each other (the nucleus spark rejoins with its opposite charge) and a tremendous burst of light and power is released (the reconnection). From this brilliance, the radiant whole and reunified souls go back to the God from whence they came, and the planet is restored to harmony.

Tears flooded my eyes when I saw this movie, for it so perfectly expresses how limited and narrow viewpoints can obscure the greater truths of those who dare to see beyond "the veil." Granted, because of societal conditioning, it is difficult to defy family approval or group consensus. And, during our span of life on earth, we individualize our egos to such a degree that no other destiny save death and darkness seems possible for us. This is why awakenings are so important—the light of enlightenment, even if only partial, helps us to wake up so we can remember who we are and why we're here.

The One Mind issuing forth from the matrix webbing of its own consciousness cradles all existence in the stirring of its thought.

Life truly is God made visible.

As we embrace this greater reality, life and its living are enhanced and improved beyond measure.

Four centuries have passed since St. Teresa of Avila, the great Spanish mystic and reformer, committed to writing the experiences that brought her to the highest degree of sanctity in the Catholic Church. Near the end of her life, she wrote, "The feeling remains that God is on the journey, too."

St. Teresa's words have special meaning for us now because the torus model of evolution implies that as existence changes, so, too, does that which contains it.

16

The Unbroken Web of Wholeness

All things are connected like the blood which unites one family.
Whatever befalls the earth befalls the sons of earth. Man did not
weave the web of life, he is merely a strand of it. Whatever he
does to the web, he does to himself.
—Chief Seattle

Each member of creation links into a network that, in turn, inter-weaves systems that interconnect grander patterns—like glistening threads in a lattice of layered intelligence—extending from the tiniest microbe to the complexity of the universe, including substratas not yet visible as physical form.

Our world is both reciprocal and synergistic (mutually related while also mutually interactive and interdependent). In other words, everything is joined. And that which is joined actively works to maintain the integrity of its wholeness.

Notice the connections and you'll recognize the web.

Since learning how to recognize connections and the mutual exchange they foster is important if we are to broaden our understanding of life and the flexibility of time and space, I've decided to demonstrate what I mean when I say "connections" by wrapping this entire chapter around some simple yet startling observations from the natural world. These examples focus on our earthly bodies and our immune systems, animal and plant interdependence, weather patterns, subjective and objective states of perception.

Alternate realities are easier to recognize once we acknowledge the multiple facets of our own reality. Nothing is hidden, only ignored. Reality, as we think it exists, depends entirely on the angle with which it is viewed for definition.

To put it another way, where we stand determines what we see.

Pesky ants were destroyed by city planners in Southern California and months later there were no butterflies. It seems the butterflies were dependent on the ants for part of their life cycle.

Dust blowing in from the Sahara nourishes the rain forests of the Amazon. Without the dust the rain forests suffer.

A tourist concessionaire set up shop in a cave near a forest of saguaro trees in Arizona. The cave turned out to be a roosting place for Sanborn bats, which cross-pollinate saguaros; ridding the cave of bats disrupted the pollination pattern for hundreds of miles around. This jeopardized future saguaro growth and reduced their numbers 75 percent.

James Lovelock, an unusually curious chemist-biologist-inventor, noticed this kind of codependence, this connection between life-forms and the world they inhabit, and he wrote two books about it, books that have sent the scientific community into an uproar. In his 1979 publication, *Gaia: A New Look at Life on Earth*, and again in 1989 with his *The Ages of Gaia: A Biography of Our Living Earth*, he establishes and builds on his premise that the earth is alive, a living breathing entity of its own.[37] He calls earth Gaia (pronounced "guy-a") after the Greek goddess who was said to be the mother of Earth.

His hypothesis states that all life on earth cooperates to maintain optimum environmental levels much the same way the human body maintains its stable core temperature whether sweating out in the Tropics or shivering at the Poles. It is the life on earth, according to Lovelock, that actually regulates temperature, oxygen levels, ocean salinity, humidity, atmospheric chemicals, and other necessities for existence on this planet. Far from any haphazard collection of flora and fauna, earth's life-forms operate as if in partnership with one giant self-controlling organism.

Through computer modeling, Lovelock uses the growth of daisies to describe what he is talking about. When the sun is weak, the darker, heat-absorbing daisies proliferate across the planet, which raises the global thermostat by trapping whatever heat is available. But when the sun is strong, the lighter, reflective daisies predominate, thus creating a protective shield against the hotter temperatures. Also, dimethyl sulfide (DMS) produced by marine plankton accumulates in the oceans and then diffuses into the atmosphere. Once airborne, it oxidizes, leaving behind a trail of sulfate particles that serve to seed the formation of clouds. With plenty of plankton, there are plenty of clouds. When plankton levels reduce, so do the clouds and the life-giving rains they carry. Cloud-cover density, then, is directly linked to the health and the well-being of tiny marine organisms.

Unlike Darwin, Lovelock is finding in his work that it is not survival of the fittest, competition, that best explains evolution, but rather cooperation for mutual benefit, synergy. Although parts of his Gaia Hypothesis remain unproven, enough has been verified to indicate that the interconnections between life-forms on earth, our biosphere, are as if interwoven by an all-pervading intelligence—with we the people an integral part of what sustains existence.

But Lovelock is not the only one noticing how everything connects with everything else. Architect Gyorgy Doczi eloquently described and illustrated repetitious patterns of order and beauty in nature, designs revealed by slicing through a head of cabbage or an orange and by examining the form of shells and butterfly wings, in his surprising book *The Power of Limits: Proportional Harmonies in Nature, Art and Architecture.*[38]

By seeking to understand Pythagoras's epigram, "Limit gives form to the limitless," Doczi discovered recorded in the silent language of patterns what is termed "golden proportions." This mathematical formula celebrates the unique relationship between two unequal parts of a whole, where the small part stands in the same proportion to the large part as the large part stands to the whole.

Golden proportions, according to Doczi, uphold harmony and beauty and give rise to moral and mutually beneficial behavior, and he clarifies his point by reminding us that the act of sharing is also part of this formula. "If we share what we have with our neighbors who have less, we gain more than if we had kept it all for ourselves. This is the paradox of sharing expressed by the words St. Paul attributed to Jesus: 'It is more blessed to give than to receive.'"

The astonishing unity he unveils from realms of nature and art repeats throughout the human body. He writes: "The hand is a microcosm mirroring the macrocosm of the body. It grows out of the wrist as the spine grows out of the sacrum, and as wings grow out of the butterfly, or as leaves and flowers grow out of their stems."

The potential inherent within form, the power of limits, describes the harmonious diversity of a creation we often take for granted, even our very bodies and our mind. Knowing this helps us to realize that interconnections are *internal* to us as well as external.

Quite the opposite from previous medical research findings, it can now be said that the immune system rivals the brain in complexity, that it can produce virtually every type of hormone the brain produces, and that it continuously exchanges messages with the brain about its experiences in the world. Immune responses are so striking, it is almost as if the cells themselves experience fear or grief or hope or joy. This indicates that the health of our bodies is *a direct manifestation* of our state of mind, and it refutes any notion that we are helpless victims of disease (the word "disease" literally means "not at ease"). Because of this finding, the new research discipline of psychoneuroimmunology (how the mind affects the immune system) is redefining the entire premise for medical intervention when treating ill patients. A good reference for this, documenting many strides in mind-body research, is Norman Cousins' best-seller, *Head First: The Biology of Hope*.[39]

Then there is orthopedic surgeon Robert O. Becker with his stunning book, *The Body Electric: Electromagnetism and the Foundation of Life*.[40]

He writes: "The more I consider the origins of medicine, the more I'm convinced that all true physicians seek the same thing. The gulf between folk therapy and our own stainless steel version is illusory. Western medicine springs from the same roots and, in the final analysis, acts through the same little-understood forces as its country cousins. Our doctors ignore this kinship at their—and worse, their patients'—peril. All worthwhile medical research and every medicine man's intuition are part of the same quest for knowledge of the same elusive healing energy."

Becker goes on to explain that had biology and bioelectricity not separated as they did nearly two centuries ago, the electrically actuated self-healing mechanisms of the human body would have described a different model, one more consistent with prevention, nutrition, exercise, lifestyle, plus each patient's physical and mental uniqueness. The function of the physician during illness would have been to assist with the natural healing abilities of the patient. As salamanders are capable of unequaled feats of self-healing and limb regeneration, he has found that humans are also capable of similar "miracles." Along this line, he has documented that children can regrow severed fingertips in about three months; fractured bones can repair *perfectly* with just the use of electrical or magnetic stimulation.

Added to the revolution occurring in today's health field is the finding that the human body is a living sundial. The pineal gland produces a hormone called melatonin in ratio to the hours of light and darkness reaching the eye. Darkness stimulates the greatest production of melatonin, especially during the long months of winter or in locations where sunlight is sporadic. The trick is to keep this hormone in balance with our need for sleep. Too much melatonin seems to bring on depression, excessive tiredness, and mood disorders. Sunshine or occasional doses of strong light perk us up to where we are again alert and fully engaged. Left to the natural course of light and darkness, the pineal gland easily regulates our healthy response to nature's cycles.

But the pineal gland is not our only keeper of biological time. We normally feel hunger about every three to four hours, with our bodies' circadian levels (rhythmic biological cycles) dipping and peaking at almost the same rhythm. Brain cells usually operate in three-hour activity cycles and dreams occur in repetitious patterns of about ninety minutes. Temperature, blood pressure, pulse, breathing, and hormonal activity all rise and fall in rhythm with Earth's daily rotation, and are affected as well by the monthly orbit of the Moon around Earth and

Earth's yearly trip around the Sun. It has been observed in laboratory research that human blood changes electrochemically when groups of sunspots pass across the Sun's center and again, quite suddenly, a few minutes before sunrise each morning. And we respond best to a twenty-*five* hour day, not twenty-four.

Dr. Susumu Ohno of the Beckman Research Institute in Duarte, California, converted the chemical formulas of living cells into musical notes. He found by doing this that the molecules from different bodily areas each had their own musical melody, some even bearing an uncanny resemblance to the works of the great composers.

"Translated into sheet music and performed on the piano, a portion of mouse ribonucleic acid, a complex genetic messenger substance, sounds like a lively waltz." He further explained that, "except for its quicker tempo, parts of the mouse RNA waltz are dead ringers for passages in Chopin's Nocturne, Opus 55, No. 1." But when he reversed the process, converting musical notes from a funeral march by Chopin into chemical equations, he discovered entire passages identical to a cancer gene found in humans.

Ohno noted from this that a healthy body creates a living symphony of musical harmony, while an unhealthy body produces dissonance similar to a dirge.

The web of interconnections and interweavings between life-forms, however, is more complex, involved, and varied than anything mentioned thus far. That's because underlying our inner and outer worlds are intercommunication systems that embrace and blend the animate with the inanimate.

Yes, you read that right. I said "*inanimate.*"

Consider the story of Michelangelo and his creation of the famous masterpiece the *Pieta*. Given the block of stone as a gift, he sat for many hours contemplating it until he finally asked the stone, "What would you like to become?" Then one evening during meditation, the stone replied by revealing to him the form of the *Pieta* within it. "The act of creation was simple," Michelangelo stated, "for all I did was remove what no longer belonged." He released what asked to be released, what was waiting inside the stone.

Also consider the story of famous architect Louis Kahn and the incredible way he combined mysticism with poetry in his structural designs. He even talked to bricks: "You say to brick, 'What do you want, brick?' And brick says to you, 'I like an arch.' And you say to brick, 'Look, I want one, too, but arches are expensive and I can use a concrete lintel . . . What do you think of that, brick?' Brick says, 'I like an arch.'" By asking buildings what *they* wanted to be, Kahn created structures of striking beauty and functional excellence. (Refer to *Between Silence and Light: Spirit in the Architecture of Louis I. Kahn*, by John Lobell).[41]

The idea of conferring with inanimate objects is ancient in principle but modern in application, as more people are beginning to discover that one can successfully communicate with existences once thought to be lifeless. Since everything in nature contains a record of how it was made and what it is for, the reasonable approach is to consult with any given material about what it wants to become *before* committing that material to a design or purpose. The "voice" that somehow seems to answer back is worth listening to.

Actually, our knowledge of the natural world is still primitive. We respond to it on the basis of our own perceptual preferences rather than trying to determine the manner it uses to respond to us.

To appreciate what I am saying here, try this: Lean against a tree and feel what it is like to be the tree leaning against you. When you walk across a floor, feel what it is like to be the floor supporting the weight of the steps you take. As you press the wall with your fingers, feel what it is like to be the wall pressing back. As the wind blows your hair, feel what it is like to stand in the way of the wind, blocking its free passage as it sweeps by.

Each member of the whole is aware on some level of the possibilities inherent within it and the role it can play; and each member knows of its own existence.

Nature does not limit intercommunications, only humans do. In nature, the seemingly "limited" is limitless.

With this in mind, look up the May 23, 1988, issue of *Newsweek* magazine.[42] The feature article for that week was entitled "How Smart Are Animals? They Know More Than You Think." In the article, research documentation was quoted that established that animals can conceptualize, plan, and think through problems in ways similar to or sometimes better than humans; some animals are known to possess photographic memories and to have active dream lives. Although research has continuously verified how psychic and intelligent animals can be, we now know they can also abstract. Contrasting communication skills account for the biggest hurdle separating closer ties between animals and humans. This is because animals understand best the language of imagery and the sensations that come from emotional overtones, while humans respond better to pinpoint details and complex thought structures that convey and record ideas. And who's to say which language is preferable, as the greatest benefits come when each learns the other's style.

The February 3, 1992, issue of the same magazine contained another surprise about the animal kingdom—they know and understand how to use herbs for their own health. For instance, a Kodiak bear will chew a certain root to make a salve, then rub the salve over its fur when injured or irritated. Chimps know what plants to eat to cure a stomachache or get rid of parasites or prevent disease. Howler Monkeys use specific herbs to determine the sex of their babies or induce abortions.

Research has also revealed that humpback whales, like humans, use rhyme. Whether to prime their memories or not, whales sing in intricate, stylized compositions, some longer than symphonic movements, and in medleys that can last up to twenty-two hours. These songs change dramatically from year to year and include obvious rhyme. No one knows the why of it, but Katharine Payne of Cornell University's Zoology Department remarked, "They change their songs like hemlines, and provide one of the nicest examples of cultural evolution that has been gathered from any species in the animal kingdom, including man."

Plants, minerals, the environment itself, all possess intelligence, intuition, feeling, and memory—plus the ability to communicate. This fact challenges us to accept the existence of worldviews other than our own, worldviews that are not less, only different.

Sometimes the question to ask is not "What is life?" but, rather, "How alive is life?"

Vernon M. Sylvest, M.D., a physician and founder of the Richmond Health and Wellness Center,[43] made a discovery about wasps that changed his thinking about the interrelatedness of the many with the whole. This is his story:

> "Back in August 1988, I was cleaning pine needles off the roof of our home (not a task I enjoy), when I encountered a wasp nest and made contact with the stinging end of a wasp. As an experiment, I immediately went into a meditative state, and reached for an experience of God's love for myself and the wasps. An image of a large wasp filled the screen of my mind, not angrily, but sternly, and it told me I was in wasp territory. I agreed to let all the wasps have their territory and not revert to my usual action of bombing them out with Raid. My pain instantly subsided as I meditated and, afterward, the sting required no treatment and did not swell, which is unusual for me with wasp stings. But I used the sting, anyway, as an excuse not to do any more cleaning on the roof.

> "Eleven months later, the roof looked like a compost pile and the gutters were completely covered and nonfunctional. My wife and I had just returned from a vacation of backpacking in the mountains so I felt in shape, energized, and ready to tackle the whole roof.

> "So up I went with rake and blower. I was not certain if the wasps were back this summer, but I was determined to do what was necessary to clean the roof. They were back and in the same area. I started cleaning the other areas first, saving theirs for last. As I moved in their direction, I approached with caution, using the rake and not the blower so as not to create much disturbance. This did not go unnoticed. Off the nest from behind the eave they came and toward me. I backed away.

> "My first thought, the old thought, was to get the Raid and blast them. Then I remembered my previous meditation

experiment of honoring my love and respect for all life. This memory guided me into a prayerful and meditative state in which I experienced my oneness and love for all creation, including the wasps. Again, the image of a large wasp filled the screen of my mind, and I found myself addressing what I perceived as the consciousness behind the wasp colony. I explained that I respected the wasps' role in the nature of things, understood their usefulness in controlling the insect population, and that I only wanted to clean off the roof and would not disturb the nest. The wasp consciousness agreed to cooperate with me and I proceeded.

"This time as my raking and also blowing approached their nest, the wasps flew off in another direction, giving me space for doing my work. Amazing, but it happened. No stings. As I cleaned other areas of the roof, I noticed more wasps and assumed there were more unseen nests. I expanded my thoughts to include them as well, and had no problems.

"The next step was removing the wire gutter covers and cleaning the gutters. This was to take me directly under the area of their nest. I started from the furthermost corner and worked my way toward them. This may sound suspenseful as I write it, but I had no fear and sensed things would go well. However, I still approached with respect to cause the least disturbance possible. I lifted the gutter cover nearest their nest. As I lifted the cover, I found on the underside another wasp nest. Strange place for a nest, I thought. I assumed it was an old one and reached out my hand and pulled it away. I noticed its connection was unusually tough, plus the nest did not have a dried-out appearance. As I turned it over and looked at the underside, I discovered it was an active nest with live larvae. Too late now, I quickly tossed it away and backed off to meditate again.

"Sure enough, the wasps came back looking for their nest. The wasp image once again filled the screen of my mind. I apologized for destroying the nest, explaining that I did not realize it was active, and, furthermore, I could not have gotten the task done without disturbing it. The wasp consciousness responded in agreement that the location was not a wise choice and that everything was in order. My assumption is that the individual expressions of this wasp consciousness would learn from the experience and make a better choice next time. The wasp consciousness was refining its wisdom. I went back to cleaning the gutter and replacing the cover. The individual expressions of the wasp consciousness went about looking for the no-longer-existing nest, but they left me alone—as if I were invisible!

"I cannot prove, not even to myself, that I was communicating with some sort of group consciousness of the wasps, although that was the way I experienced it, and that is my feeling about the experience. The important lesson in this may be that if we release fear and feel appreciation and respect for all life, we will have that appreciation and respect returned—and our life events will be harmless to ourselves and others."

We humans are in league with countless members of a universal lifestream. What is true for one aspect of this lifestream is true for another, as each reflects the other.

Ponder this: A field of crops not in proper nutritional balance will emit a sound, much like crying, which will attract to it the very insects, bacteria, and substances needed for its restoration or destruction. A human being not in proper balance physically, emotionally, mentally, and spiritually will automatically set up a vibrational "signal" that will attract to him or her the very diseases, accidents, or episodes necessary for that individual's redirection, rebirth, or death.

And this: As we have learned in agriculture that too much and too heavy a dose of chemicals does more harm than good, so, too, we are learning that the same thing applies in medical care. To understand this, realize that only 1 percent of the pesticides applied to plants ever reach their destination. The other 99 percent pollutes air, soil, and water. With medicine, the pharmaceutical companies finance all major medical journals, aggressively court new doctors in medical school with varied enticements, send their "detail people" on a regular basis into doctors' offices to "educate" doctors on the latest drugs while pushing drug use, even though research has demonstrated over and over again that people are becoming less and less responsive to most of the drugs they are pitching. In any other profession, the business practice of pharmaceutical companies would be labeled "conflict of interest" and would be subject to court proceedings.

And this: The physicist Ed Wagner has discovered evidence that trees talk to each other in a language he calls W-waves. "If you chop into a tree, you can see that adjacent trees put out an electrical pulse," he explained. "This indicates that they communicate directly. It puts out a tremendous cry of alarm. The adjacent trees put out smaller ones. People have known there was communication between trees for several years, but they've explained it by the chemicals trees produce. I think the real communication is much quicker and more dramatic than that. These trees know within a few seconds what is happening."

And this: Cleve Backster, a pioneer in the development and application of the polygraph technique for researching cellular communication, has established that an unfertilized egg will register the heartbeat of a three-day-old embryo, as a way of preparing a protective energy shield should fertilization occur. He has also found that every cell responds to any form of consciousness, and that all cells are "notified" in advance of what is about to happen through *nontiming* (a term Backster coined to explain the instantaneousness of cellular signals).[44]

Ponder this as well: Measurements from remote sensing satellites and field observations confirm that fluctuations in cattle grazing patterns can alter local climate to a degree that exceeds the effects of greenhouse warming or urbanization. Studies of various climatology records—when compared to a given area, the people there, and historical or current events—suggest a possible link between weather fluctuations and fluctuating patterns of attitude and behavior from local residents. (Even though the necessary large-scale research to prove or deny this possible link has yet to be done, the activation formula of one-tenth of 1 percent from the law of resonance points to what may be an explanation. In chapter 8, we explored how it doesn't take much in the way of dominant or focused energy to initiate the resonance of relationships in motion. This explains how a single individual could indeed sway the masses; yet it also addresses how small numbers of people could alter not only the politics in their part of a given country, but perhaps even the very weather they live in—including incidents of earth movement. The principle that activates the power of prayer appears to operate via the same formula.)

Thus, responsible behavior seems tandem to "membership" in the universal lifestream if mutual benefit is to be ongoing. Here are some examples of what happens when we the people ignore this responsibility: Acid rain caused by power stations in Great Britain kills trees, even threatening whole forests, in Scandinavia. The hamburger craze in North America causes tropical forests in South America to be destroyed so more cattle can be raised to supply more meat for more hamburgers. Both areas of tree reduction directly affect the oxygen levels necessary for healthy life on this planet.

We are each part of life's continuum; interrelated, interdependent, interconnected, and symbiotically attached. We are each part of an unbroken web of wholeness. While short-range planning usually leads to destruction and entropy, long-range planning almost always stimulates innovation and renewal. As the Native American tradition cautions, make all plans to the seventh generation.

The unbroken web of wholeness includes our perception of time's passage, as well. One can learn to recognize through observation how the energy undergirding each moment *mirrors what happens within it*—since events and participants are imprinted with qualities unique to that moment's energy. (This is the premise most psychic impressions are based upon, being able to "read" or understand the imprint of energy at any given time. Astrology operates from a similar premise.) To interpret the possible meanings of these qualities, I have learned to be alert for whatever "signals" catch my attention. This enables life's mirror, the energy patterning of the moment, to be revealed.

To illustrate how time's passage can mirror the energy within it while imprinting qualities that become as "signs" that can be interpreted, here is a particular incident I witnessed in 1979. As you read this, notice how each moment reflected back the energy of what was *really* going on, and this was done through seemingly unrelated but corollary images and activities:

> March the twenty-sixth on the North Lawn of the White House, the Peace Accord was to be signed. President Carter, Menachem Begin, and Anwar Sadat would participate in public ceremony.
>
> I managed to rearrange my lunch hour and scoot out the door, arriving just in time to hear a woman's voice announce the beginnings of the ceremony. What a nice touch, I mused, at least someone is giving a woman an opportunity to do something of merit. I found space to stand atop one end of a bench in Lafayette Park, hardly a stone's throw from the street and in full view of the proceedings. Throngs quickly packed the area shoulder to shoulder. Skies were heavy and gray and the wind was coldly fierce. Arab protesters nearby shouted defiantly, but no one paid them much mind. What mattered, all that mattered, was those three men standing in front of the White House and what they were about to do.
>
> Carter spoke first, but his voice somehow fought the loudspeakers, resulting in little more than a garbled ramble. Sadat was next. With his first word, all the birds in Lafayette Park took flight en masse, flew to the area where he stood, and landed. A sudden splash of warm sunlight broke through the clouds and followed the birds, pausing long enough to illumine Sadat and the area around him. When he spoke, Sadat's voice was clear, smooth, and easy to understand, his message of universal oneness inspiring, almost reverent. When Begin took his turn, the birds again lifted en masse and returned to the Park. The sun retreated, leaving the sky as gray and cold as before. Begin's voice, like Carter's, faded in

and out of the microphone, his words almost impossible to understand until his final shout—"Shalom, peace."

The Accord was signed. The bells of St. John's Church rang joyously and the crowd cheered. It was a moment in history, and I felt privileged to be there. Still, nature had made known, at least for those willing to "see," that Sadat was the real hero of this event. Because of how the birds celebrated his being, how the moment honored his light with the grandeur of the sun, and how his voice rang with such purity, these signs showed me that Sadat would pay for his signature on that document with his life; all else but him remained wrapped in grayness, unable to appreciate the fullness of his vision, A tear fell, for there was nothing I or anyone else could do to change the outflow of events. I could only accept what Sadat himself must have already known on some level of his being, that the greater glory of this special moment needed to be.

If you find yourself astonished by the volume of material presented in this one chapter, then I've done my job.

If you find yourself utterly overcome by the type of material presented throughout the entire middle section of this book, then I've made my point: Most people are aware of only 1 percent of their inner and outer worlds. We miss 99 percent of what's really there. If we could but enlarge our perceptual viewing angle, we would discover an "imprisoned splendor" of awesome dimensions.

That which exists is not just interconnected, nor is it just in active dialogue with itself; it is part of an ongoing flow of intelligence that, like future memory, demonstrates the dynamic power interspersed throughout creation and resident within the lives and consciousness of each and every one of us.

Part III

Beyond Illusion

17

The Labyrinth Revealed

*The consciousness of each of us is evolution
looking at itself and reflecting upon itself.*
—Teilhard de Chardin

Anyone can suddenly begin to experience future memory episodes, even without having undergone a life-changing transformation.

Most future memory episodes center around daily, mundane occurrences. Seldom is anything major touched upon, yet invariably, experiencers comment on how their life seems to flow more easily and naturally after having had one. Although it is spontaneous, there is an obvious pattern of growth present as if these individuals were developing "mental muscles" and becoming more perceptive while at the same time stretching to reach greater dimensions within their minds. Sometimes, once an individual becomes comfortable living in this manner, future memory incidents fade—but not always. The phenomenon can increase in frequency and so can the sensitivity it engenders, enabling individuals to access heightened states of clarity more often and more reliably.

Perhaps future memory is not so much a curious aftereffect of transformative experiences as a physiological side effect from the process of brain shift.

Certainly, a brain shift is what usually happens to people who undergo a spiritual transformation, religious conversion, near-death episode, shamanistic vision quest, kundalini breakthrough, certain types of head trauma, or who have been hit by lightning (turbulent method). But it can also come about from the slow, steady application of spiritual disciplines, mindfulness techniques, or because, in a prayerful state of mind, an individual simply desires to become a better person (tranquil method).

You can tell if someone is in the process of shifting, or has had a partial or full brain shift, by the type of behavior adjustments made during and afterward. Here is a profile of the pattern universal to the experience (refer to my previous book, *Beyond the Light,* for a more detailed description).

Major Characteristics Displayed by People Who Have Gone through a Brain Shift

Physiological: Changes in thought-processing (switch from sequential/ selective thinking to clustered thinking and an acceptance of ambiguity), insatiable curiosity, heightened intelligence, more creative and inventive, unusual sensitivity to light and sound, substantially more or less energy (even energy surges, oft-times more sexual), reversal of body clock, lower blood pressure, accelerated metabolic and substance absorption rates (decreased tolerance of pharmaceuticals and chemically treated products), electrical sensitivity, synesthesia (multiple sensing), a preference for more vegetables and grains (less of meat), physically younger looking (before and after photos can differ).

Psychological: Loss of the fear of death, more spiritual/less religious, abstract easily, philosophical, can go through bouts of depression, disregard for time, more generous and charitable, form expansive concepts of love while at the same time challenged to initiate and maintain satisfying relationships, "inner child" issues exaggerate, less competitive, convinced of a life purpose, rejection of previous limitations and norms, heightened sensations of taste-touch-texture-smell, increased psychic ability and future memory episodes, charismatic, childlike sense of wonder and joy, less stressed, more detached and objective (dissociation), "merge" easily (absorbtion), hunger for knowledge and learning.

Note: Characteristics can be positive or negative, depending on application. With my research of near-death survivors, I found the spread of impact as follows (1994 figures): 21 percent claimed no discernible changes afterward, 60 percent reported significant changes, and 19 percent said changes were so radical they felt as if they had become another person. This range of percentages seems to fit across the board with the universal experience of a brain shift, no matter how caused.

We discussed earlier the similarities between people who have gone through a partial or full brain shift and youngsters at about the age of three to five. Now let's do a more specific comparison.

BRAIN DEVELOPMENT COMPARISON BETWEEN THREE-TO-FIVE-YEAR-OLDS & BRAIN SHIFT EXPERIENCERS	
THREE-TO-FIVE-YEAR-OLDS	**BRAIN SHIFT EXPERIENCERS**
Temporal lobe development *Emerging consciousness*	*Temporal lobe expansion* *Enlarging consciousness*
Prelive the future on a regular basis; spend more time in future than in present.	Prelive the future on a regular basis through dream states, visions, future memory episodes.
Play with futuristic possibilities as a way of "getting ready"; rehearse in advance demands soon to be made upon them.	Preexperience life's challenges and opportunities before they occur as a way of preparing to meet futuristic demands.
No natural understanding of time-space states; consider "future" an aspect of "now." Gain perspective and continuity by establishing the validity of "future" (continuous scenery and connected wholes).	No longer restricted by a sense of time-space states; an awareness of simultaneity and the importance of "now." Embrace broader dimensions of experience beyond that of "future" (unlimited perspective).
Progress from archetypical mental models to stereotypical ones in a process of self-discovery.	Progress from stereotypical mental models to the process of individuation in a journey of soul discovery.
The birth of imagination	The rebirth of imagination

The "Brain Development Comparison Between Three-to-Five-Year-Olds & Brain Shift Experiencers" chart highlights how reliable the future memory phenomenon is as a signal that a person's brain is shifting in structure, chemistry, and function. I have also made the following observations:

- Being able to live the future in advance, and remembering that one did, alleviates much of the stress and fear that worrying about unknown variables can cause. This advance "rehearsal" enables the human psyche to negotiate the demands of sudden change more smoothly. The ability imparts an immense sense of confidence and peace in individuals, no matter what age, and often leads to frequent internal and external flow states.

- Preliving the future has less to do with "psychic" forms of futuristic awareness than it does with the development of *the higher brain*—so *the higher mind* can be successfully accessed and utilized. As more and more people expand in consciousness, future memory episodes will become widespread.

There are three evolutionary stages of brain development, classified as primitive brain, midbrain, and higher brain. Such a powerful change in one's consciousness as a brain shift is known to accelerate higher brain activity. One progresses (evolves) into and through this final stage of brain development as any child would, and for the same purpose—*maturity*. Only, in this case, that maturity is a synergistic type of wholeness gained from heightened levels of intelligence and knowing.

Research has shown that the brains of children who are deprived of ample stimulation during the first few years of life suffer some degree of retardation—sometimes permanent brain damage. Adults are stagnated as well if they are denied or avoid or ignore challenge. Growth spurts in a child's brain, though, are cyclic, heralding specific stages where development can advance. As the brain ages, and in tandem with a natural slowing in the spread of nerve networks, the brain itself is still quite capable of increasing intellect; but growth spurts on the magnitude of what one finds with children are rare.

I have noticed that a brain shift seems to jump-start the brain's capacity and capability. The growth spurt that appears to result overrides whatever level of development is present and triggers great swells of vitality plus an insatiable desire to learn.

Internal imagery played out in the mind during an actual brain shift is experienced as an "otherworld journey." For this, our psyche seems to fashion from our hangups, urges, desires, and longings, or from collective stereotypes or archetypes, the basic form and shape of the story line our journey takes, which, if truth be known, is but a "stepping down" of the colloidal condition to a vibrational frequency and image patterning we can respond to. Hence, the presence of such things as angels, light beings, aliens, pathways, pillars, libraries of knowledge, and so forth is more commonly reported than any impersonal geometries or abstracts that might challenge our ability to comprehend. (An example of this is the illusion television creates: It's the story line we respond to and learn from, *not* the reality of what's really there, that is, rolling raster bars and an electron "gun" shooting tiny dots at a plate of glass.)

If the shift is profoundly deep and involved, there is a chance that image patterning will leap past both personal and collective thoughtforms to reveal the levels of consciousness where Truth resides—a realm of indescribable luminescence. (Along this line, of the thousands of near-death survivors I have interviewed, *in each case* where the individual asked the angel or godlike figure what he or she really looked like, that winged one or fatherly type on a giant throne instantly dissolved into either a power burst, into geometries such as a sphere or cylinder of light, or into a pure sine wave.)

The colloidal condition suspends time-space states, then expands, enlarges, imprints, and transmutes, in tandem with a power surge; in the brain, it's a growth spurt (that's how you recognize that a brain shift has happened).

A brief review of our cultural understanding of time and how the concept of "future" has altered, plus how that mimics the three stages of brain development, will help us to make an important observation:

- Kairos, the kairotic (chaotic) and leisurely flow of a more organic rhythm that springs from an inner cadence—a response to nature's cycles (future based on feelings; primitive brain).

- Chronos, the chronological progression of that which is linear, orderly, quantifiable, and mechanical—a response to calendars and clocks (future based on intellect; midbrain).

- Computime, the universal alignment in cyberspace to nanoseconds (a billionth of a second), virtual simultaneity—a response to global computer grids (future based on abstract; higher brain).

The observation?

Contrast.

Remember, mind is nonlocal and consciousness is a field effect. Mind is primary; consciousness, the space that cradles light and gives it being. Time and space merge and become simultaneous when nonlocal (primary). You gain a state of oneness when this happens, but you lose contrast.

To grow, to experience and learn and recognize and respond, to know thyself, even to develop in fullness as the souls we are—*there must be contrast.*

Our cultural understanding of time and the concept of "future" explains this. Contrast on the earthplane cannot exist in any form without organic time (*kairos*), cannot recognize the value of self-determination without linearly ordered time (*chronos*), cannot evolve into larger, grander orders without the simultaneity of abstracted time (*computime* or heightened states of enlightenment).

Future is what guarantees the progression of life as we know it.

Future enables perception to develop and the continuity of form and shape to be established and maintained.

Future is what instills the worth of effort and activity, while accommodating the spread and the growth of consciousness as it contrasts internal and external awareness potentials.

Without timing and a sense of future, life on the earthplane would be utterly meaningless.

The hinge is contrast.

The key is future.

Once the importance of future is understood, the purpose of existence—life and death and memory and matter and creation and consciousness—begins to make sense. At last, the labyrinth can reveal its deepest secrets.

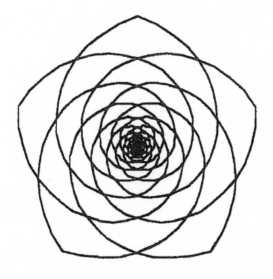

A strand of DNA as viewed from the top down.

18

Fixed and Flexible Futures

Few men have imagination enough for reality.
—Johann Wolfgang Von Goethe

Many people have premonitions of a future that later happens. Historical examples:

- During the worst days of the Revolutionary War, when despair and defeat loomed heavy, George Washington went into his quarters at Valley Forge and did not emerge for hours. When he did, he seemed aglow and filled with the courage he needed to turn the tide of battle. His aide, Anthony Sherman, recorded Washington's exact words when he stepped from his tent—to tell of a visitation by angelic hosts who had shown him the birth, progress, and destiny of the United States.[45]

- Less than two weeks before his death, Abraham Lincoln had a disturbing dream. In it, he heard sobs and great weeping. As he wandered through the various scenes in his dream, he found himself in the East Room of the Capitol and there saw a catafalque on which rested a corpse wrapped in funeral vestments. Soldiers acting as guards stood at attention amidst the grieving throngs. He demanded of one of the soldiers, "Who is dead?" "The president, killed by an assassin!" was the reply.[46]

- The morning of Lincoln's assassination, the wife of General Ulysses S. Grant awoke consumed with fear, knowing that she, her husband, and her child must leave for their home in New Jersey immediately—even though they were to accompany the Lincolns to the theater that night and sit with them. She badgered her husband until he finally agreed to go. The dire news reached them as they were traveling. Later they learned that the general was on the assassin's list of intended victims.[47]

- In January 1967, United Press reported that a few weeks before the fiery disaster of Apollo 1, which killed astronauts Gus Grissom, Roger Chaffee, and Edward White, Grissom was discussing the risks of space flight with reporters when he suddenly turned serious and said, "We hope that if anything happens to us it will not delay the program."

- In his 1982 book, *The Cosmic Code*, noted physicist and mountain climber Heinz R. Pagels wrote, "I dreamed I was clutching at the face of a rock but it would not hold. Gravel gave way. I grasped for a shrub, but it pulled loose, and in cold terror I fell into the abyss." He went on, "What I embody, the principle of life, cannot be destroyed. It is written into the cosmic code, the order of the universe. As I continued to fall in the dark void, embraced by the vault of the heavens, I sang to the beauty of the stars and made my peace with the darkness."[48] Six years later he died as stated in his book.

Some people encounter peculiarities of future whereby time overlaps itself. Case in point, the story of hypnotist Dolores Cannon.

In the early 1970s, Cannon developed a unique method for discerning information from people she hypnotically regressed into past lives, and in so doing, she happened upon the personage of Nostradamus. It made no difference whom she hypnotized, as she used a mirror she could "see" in his study for a focal point in her communications with him. Once contact was made with Nostradamus, conversation began where it had previously ended. Always, though, she had an eerie feeling that he was quite alive during his own time frame in France when they linked together, and that somehow it was she who was supplying him with the material he used to write his quatrains. Years later, she read in one of his biographies that he claimed his revelations came from "spirits of future" who spoke to him through a mirror in his study.

When I interviewed her at a meeting in New York City, Cannon offered this explanation: "If he were alive at the same time we are alive, that brings up the theory of simultaneous time, which is something I do not understand or know if I wish to understand. Every time I tried to think of this concept of everything occurring at once—past, present, and future—I did not feel enlightened. I just got dizzy."[49]

Was Dolores Cannon the source of Nostradamus's foreknowledge? Had she inadvertently discovered a way to bridge the chasm of time so past, present, and future could merge? Is simultaneity valid as a physical reality?

Stephen Hawking in his book, *A Brief History of Time*, introduces a concept called "imaginary time," where imaginary numbers can replace real ones as a way to measure time. By doing this, he allows the distinction between time and space to disappear completely so there are no differences in forward and backward time lines—one can go in either direction depending on circumstances. According to Hawking: "This might suggest that the so-called imaginary time is real time, and that what we call time is just a figment of our imaginations. In real time, the universe has a beginning and an end at singularities that form a boundary to space-time and at which the laws of science break down. But in imaginary time, there are no singularities or boundaries. So, maybe what we call imaginary time is really more basic, and what we call real is just an idea that we invent to help us describe what we think the universe is like."[50]

Hawking believes that our ability to remember the past but not the future is based on entropy (the rate at which energy is no longer available for use and becomes inert). He reasons that whenever a memory is made, be it in a brain or a computer, the tiny amount of energy needed to activate necessary neurons or electrons is released as heat. Heat once released increases entropy, and entropy increases from yesterday to tomorrow—thus establishing the arrow of time. The idea that one can only remember what has already happened does not explain how time functions, Hawking claims, but the dissipation of heat does.

What Hawking hasn't considered here is the suspension of time and space in the colloidal state. Heat is contained, then accelerated. When this occurs, nothing is released, only changed. To test this, examine what happens in nature when a particle goes through a colloidal condition. It doesn't lessen; it enlarges and enhances and becomes permanently imprinted by a transmutation process that reverses entropy. Those who go through a brain shift (also a colloidal state) experience the same thing as they appear to roll back the clock and start life over—younger, smarter, wiser, physically and psychologically altered (to the degree of the shift they had).

To sum up what has just been said: The release of heat creates the arrow of time from past to future; the containment, acceleration, and transmutation of heat creates simultaneity—the enfoldment of time and the possible reversal of entropy.

To put this in perspective, let's take a look at time and space again. Science tells us that time and space are properties of energy. "Time" is thought to be created by the oscillation of energy as it vibrates (and is based on heat release factors, according to Hawking); "space," by the length of measurement between opposing poles of attraction set up by that oscillation. Time, at its core essence, then, has nothing to do with clocks and wristwatches and calendars; neither does space relate to distances spread out across our planet or through the sky above us. Time is reversible; circumstance, its only determinant.

Time and space are but concepts relative to the perception of the perceiver. As Hawking is fond of saying, "Each observer has his own measure of time."

The logic of subatomic physics reveals countless worlds thriving alongside our own, invisible and beyond ordinary reach. This new science indicates that we are not where we are by accident, that we have by our very existence chosen the world we inhabit from among a vast array of other choices, and that future and our memory of future can be accessed as any other memory because all events exist in the realm of thought-forms (the implicate frequency domain) and can coalesce into matter when conditions are right. (This thought-form realm is commonly referred to as the "blueprint level" of manifestation.)

Beyond what seems to be reality are realities without number.

Yet the illusion of dependability (time and space) and of stable substance (matter) serves an important purpose. To understand why, consider electrons.

The "actual" electron is stable, like true north on a compass, but the "virtual" electron can fluctuate, like magnetic north on a compass. The former presents a model considered "true" because it is long term and consistently dependable, while the latter offers an "approximation" based on its short-term capacity for maneuvering. The short-term version could not grow or change or alter without the long-term's stability, yet the long term needs the short term in order to experience itself.

That which is true supports the appearance of that which seems to be true so the necessary flexibility for growth and change (contrast) can exist.

Variation is guaranteed because that which seems real, isn't.

Actual and virtual electrons, why they exist and how they function, offer a persuasive model of how the perceptual biases of living beings can and ofttimes do overlay preexistent information to shape and form the world that appears to exist.

Think about this for a moment.

As long as the world around us appears to be what we think it is, we have all the "elbow room" we need from time and space and matter to learn and grow and evolve (like with the virtual electron). We would be lost in a meaningless jumble if we could actually see everything that exists *as it really exists* (the actual electron) before we are ready to.

Remember the perceptual illusions discussed in chapter 10? No television show could be deciphered or enjoyed if we saw everything that happened on the glass plate. No movie could have a continuous story line or plot if we saw each movie frame and the darkness in between the frames as separate units. Nor could we be emotionally affected by music or speech if all we ever heard was one sound at a time separated by silence (this would even negate effective communication).

Watch children.

If you cram their world with too much too soon, overwhelm them, they will either ignore the situation, rebel, or go into shock. But if you give them a little at a time, introduce things slowly, not only can they handle just about any challenge, they will ask for more. When provided "safe haven," some type of reliable security and dependable stability, children successfully integrate their life experiences and ready themselves for their goal of adult maturity.

All of us become as children when we begin each new progression from one level of experience to another, from one embodiment to another, from one plane of existence to another, until we successfully integrate our experiences in preparation for our goal of soul maturity.

It has been my experience that we graduate from each stage of our development in celebration of the next new beginning. The more we learn, the more we can learn, until we are able not only to see existence as it really is but to comprehend its meaning and purpose. We complete our many graduations when we enfold back into The All from whence we came. Reunion is reidentification with The Source We Are.

But, as indicated by the truth undergirding the illusion of time and space, we never really seem to "go" anywhere or "do" anything when we progress, for the only movement that actually moves is the cosmic breath *that the great thought stirs.* We, as "potential in mind," appear to progress because our thought waves (conditions of potential) behave as separate and distinct manifestations, when, I suspect, they are really interconnected fluctuations from the same source.

Suggested here is the very real possibility that the future has already happened and we are an afterthought.

Hypothetically then, as long as the centerpoint continuously reprocesses and reissues structure, the illusion of time/space/matter can maintain the continuity needed so that, through variation (contrast), we can develop our potential

Both fixed and flexible futures are real and exist at *the same time*—because that which has already happened, is still happening!

Thus, the idea of past, present, and future is simply a description of how contrast is perceived.

A model of the presentation just given is on the following page.

In the book *The Looking Glass Universe*, by John P. Briggs and F. David Peat, the authors, while discussing the work of physicist David Bohm, note this fascinating fact about the role of perception:

> Psychologists have shown that our perceptual apparatus (eyes, ears, etc.) abstract relatively unchanging or "invariant" features from the environment to create mental maps.

A map of a highway doesn't show the pavement with its changing potholes and patches or the shrubbery along the way, but only abstracts (pulls out) certain features such as the curves and direction of the roadway. The mechanisms of memory and perception do the same thing. They store and react to relatively "invariant" aspects. Once a mental map is formed, it then conditions further perception.[51]

A MODEL OF HOW EXISTENCE COULD EXIST		
Existence	Actual electron	Preexistent information from the great thought which stirred, evenly distributed throughout the universe in the form of circular, microwave logic units and containing zero-point energy. *(Original consciousness)*
Reality	Virtual Electron	Preexistent information, as "potentials in mind/" overlaying itself in countless multiples for the purpose of expression, fluctuation, variation, growth, change, contrast. *(Adaptations of consciousness)*
Centerpoint	Colloidal condition	The centerpoint of the torus gyroscope, where all structures, forms and processes from preexistent information are recycled for reuse, then integrated/ stabilized into the whole. *(Evolution of consciousness)*

With that fact duly noted, consider this statement from neuroscientist and Nobel Laureate Roger Sperry:

> Current concepts of the mind-brain relation involve a direct break with the long-established materialist and behaviorist doctrine that has dominated neuroscience for many decades. Instead of renouncing or ignoring consciousness, the new interpretation gives full recognition to the primacy of inner conscious awareness as a causal reality.

What Sperry has recognized, other researchers have too. Reality, as we think it exists, is not so much a static state as it is a process of potentiality continuously expanding and transforming so that information can increase.

Information!

Everything that exists is encoded with information (preexistent, circular microwave logic units). It seems to me that what grows, the only thing that really changes, is information expressing itself (the great thought as it stirs).

On the mundane level, our awareness of a "destiny" prodding us along does not negate our ability to alter the part we play in The Greater Plan. The extent to which we can deviate depends on how alert we are to what is happening around and within us, and how willing we are to coparticipate and cocreate with our greater power surge (soul)—rather than just coexist.

We can change the future (alter the ripples of thought waves), or we can let it happen (leave the ripples of thought waves alone). Either way, the choice is ours.

Free will is our birthright because the potential for contrast guarantees variation.

According to our actual/virtual electron model of how existence could exist, we can see how volition is vouchsafed by the illusion of time and space and matter. There could be no free will, no variation, no change, no growth, no choice . . . without something like the virtual electron, an accommodation within the structure of creation, that would allow alternate realities to overlay preexistent form.

Contrast is the hinge.

Future is the key.

The Bible says we are created in the image of God. What better way for *us,* as thought-forms (potentials in mind), to discover our true identity and the real source of our true power than through the interplay of vibrational harmonics provided by creation's magnificent illusions. Our parent image, that which we seek, is The Great Thought We Are. By allowing contrast (free will), God has given Its Own Thought the ability to independently experience Itself.

After a transformation of consciousness, one is better able to recognize this—the difference between appearances (what seems to be real) and truth (what really is real)—because one's focus of attention changes. One loses the necessity for appearances to match perceptions, and one gains the freedom to enjoy multiple and less restrictive forms of awareness. This expansion of focus enables the individual to perceive any form of reality (appearances) as but a backdrop on the stage of life. The trick, experiencers will tell you, is, "Don't identify with the scenery or the script"; remember always Who Is Really Running The Show.

The phenomenon of future memory is one of many ways we can tap into the glorious tapestry of unbroken wholeness and glimpse simultaneous-everywhere information. We remember the future because our true future is in remembering who and what we really are, and then behaving accordingly. Once so "reborn," regardless of how, we are called upon to return to society not as reformers but as transformers, preparing the way so others can awaken as well.

The phenomenon of future memory aids individuals in recognizing and processing multiple futures without the excessive tension sensory overload can cause. Operating as a shock absorber, the phenomenon signals the development of the higher brain.

19

The Higher Brain

The gravest events dawn with no more noise than the morning star
makes in rising. All great developments complete themselves in the
world, and modestly wait in silence, presenting themselves never, and
announcing themselves not at all. We must be sensitive and sensible if
we would see the beginnings and endings of great things.
—Henry Ward Beecher

Back to Lincoln.

There is evidence to suggest that he may have undergone a brain shift during his youth, perhaps several.[52]

When a child of five, Lincoln fell in a rain-swollen creek and drowned. His older friend Austin Gollaher grabbed his body and, once ashore, "pounded on him in good earnest." Water poured from his mouth as he thrashed back to attention. Although there is no record of the young boy's confiding an otherworld journey like the near-death experience to anyone, ample remarks were made by friends and family observing how suddenly he developed a craving for knowledge afterward, insisted on learning to read, and went to exhaustive lengths to consume every book he could find. Five years later, just after his mother's death and before his father remarried, he was on a wagon driving a horse and yelled "Git up," when the horse kicked him in the head. He hovered at death's door throughout the night, with his sister Sarah in attendance. On reviving, he completed the epithet aimed at the horse, ". . . you old hussy." Little more can be gleaned about the incident until, as an adult and speaking in third person, he is quoted as saying, "A mystery of the human mind. In his tenth year, he was kicked by a horse, and apparently killed for a time."

However, if you search through the numerous recollections and letters about him, there emerges a distinctive pattern of behavior changes that links back to his childhood bouts with death—a pattern typical of one who had had a full-blown brain shift that probably was triggered by a near-death experience. This observation is reasonable, since thanks to Melvin Morse, M.D., and associates, who conducted a clinical study of children and the near-death phenomenon,[53] it can now be said that about 70 percent of those youngsters who brush death, nearly die, or who revive after death has occurred *do have* a near-death episode. But tiny ones lack language skills to describe what happened to them. Ofttimes, by the time they are of school age, they either have forgotten the event or refuse to speak of it. When addressing the question, "Did they have one or not?", look for the all-important clues of a possible growth spurt in the brain above and beyond what might be expected at their age level.

Among the characteristics suggestive of a brain shift that Abraham Lincoln came to display: the loss of the fear of death, a love of music and solitude, unusual sensitivity to sound and light and food, sensing in multiples, wildly prolific psychic abilities, a preference for mysticism over religion, absorption tendencies (merging with), dissociative tendencies (detaching from), susceptibility to depression and moodiness, increased allergies, regular future memory episodes, hauntingly accurate visions, the ability to abstract and concentrate intensely, clustered thinking, charisma, moral upliftment, brilliant mind, perseverance in the face of problems or obstacles, and a driving passion about his destiny in life.

Certainly, the argument can be made that Lincoln's many idiosyncrasies were the result of his extreme poverty as a youth, coupled with a relentless determination to succeed. Yet, nothing during his early years indicated genius; *none* of his unusual talents appeared until *after* he had survived two close brushes with death. As an adult, he nearly died again and, once more, exhibited signs that he possibly could have had another acceleration in brain development triggered by another near-death event.

The same thing could be said of Albert Einstein.[54]

At the age of five, he nearly died of a serious illness. While still sick abed, his father showed him a pocket compass. The fact that the iron needle always pointed in the same direction no matter how turned impressed him that something existent in empty space must be influencing it. Although speech fluency did not occur until around the age of ten (perhaps because of dyslexia), family members recall how deeply he would reflect before answering any question—a trait that made him appear subnormal. Interestingly, he learned to play the violin at six (later delighting with the mathematical structure of music), taught himself calculus at fourteen, and enrolled in a Zurich university at fifteen. Like Lincoln, he was plagued with nervousness and stomach problems and nearly died from these afflictions as an adult, and also like Lincoln, the unusual characteristics of his temperament and talent trace back to the age of five and afterward.

And Mozart.[55]

Even though he seemed born into musical genius, there was a noticeable leap in his abilities after hovering near death's edge at the age of six, when he was mistakenly diagnosed as having scarlet fever, again at seven when he was seized with rheumatic pains in his joints, and once more when he barely survived a bout of typhus at eight. He went on to develop the same sensitivities and characteristics evident of a brain shift as did Lincoln and Einstein, and the same health challenges as an adult—until succumbing at thirty-five.

And Edward de Vere, the seventeenth earl of Oxford.

The direct connections between episodes of nearly dying as a child and then displaying enormous jumps in brain development after each event (even though already recognized as a child prodigy) are nothing short of miraculous in de Vere's case. More startling is how each spurt *exactly* dovetailed the creation of "Shakespeare's writings." Many believe that de Vere is the true author of Shakespeare's works, which expanded the English language by some 3,200 words, originated "theater" as we know it today, and gave us the concept of patriotism. So remarkable are these connections that I have devoted appendix V to a brief outline of de Vere's life and how that relates to what is known about Shakespearian authorship. This amazing material comes from the noted historian Leslie Anne Dixon, who is herself an experiencer of multiple near-death episodes—each followed by the now familiar pattern of brain shift behavior changes.

I have discovered that many of the most creative scientists, mathematicians (especially physicists), musicians, artists, inventors, and even the best psychics, both past and present, exhibited this same jump in brain development either after a near-death experience (usually as a child) or from some other type of encounter on the order of a mystical or spiritual breakthrough. Each one, in ways unique to that individual and the times in which he or she lived, came to recognize the difference between reality and illusion and made a significant contribution to society as a result.

It seems plain enough that brain shifts accelerate higher brain development. And any discussion of brain shift brings us back to the limbic system.

Simply put, if the limbic system isn't jump-started, there is no brain shift. The limbic system, as the higher brain's emissary, is the region of the brain that activates the surges in neural network expansion. Accessing mechanisms flourish in such a climate, throwing the system first into chaotic disarray until, through pathways newly formed, coherence results.

True coherence (regardless of circumstances) does not come from sticking to the course of least resistance (what seems compatible and sensible), but by allowing the novelty of ever increasing diversity an opportunity to seek the ratio of its own balance.

It is interesting to note that every religion that has ever existed has in its core teachings the very steps that lead one into the colloidal states that produce brain shifts. These core teachings are mystical, not dogmatic. However, techniques, like the one created by James Van Avery to prelive future events, quicken limbic involvement even further and take higher brain development into the population mainstream. And this needs to happen.

Why?

Because the higher brain can outwit even the best supercomputer.

And because further development of this evolutionary brain extension could well become a matter of human survival, especially by the year

2012. It has been estimated by some scientists that by the first six months of 2012, global information could double each day. During the second six months of that year, they claim that global information could double each hour. If these scientists are right, come 2013 the doubling of global information could well be *each and every second.*

Such a gargantuan onslaught could render obsolete all manner of record keeping and data storage and retrieval, since our present systems are designed for finite, not infinite, sources of information. Already, fiber-optic lines are capable of transferring an entire twenty-volume encyclo-pedia from one end of the United States to the other, via telephone with modem and computer attached, in barely one second.

Past history documents that, with each doubling of global informa-tion, unexpected revolutions occur (both violent and nonviolent) that radically change the world.

But this type of unpredictability can be found in any system where information suddenly increases and chaos follows (as with brain shifts). Yet, as order disintegrates into chaos, *that very chaos gives birth to a new order.* (Excellent resources about the science of chaos are *Chaos: Making a New Science*, by James Gleick; *Nature's Chaos*, by James Gleick and Eliot Porter; and *A Turbulent Mirror: An Illustrated Guide to the Chaos Theory and the Science of Wholeness*, by John Briggs and F. David Peat.)[56]

The law of chaos directs most of the familiar processes in the every-day world around us, from our heartbeat and thoughts to the formation of clouds and storms, from the spread of a forest fire to the path taken by a winding coastline. This law illustrates how the irregular becomes regu-lar again, how patterns throughout existence repeat themselves on large and small scales, and how *life utilizes random unpredictability to ensure continued advancement.*

Some examples of chaos:

- The focused power of coherent light emerges only from separate and diverse light waves oscillating at thousands of different frequencies, which are *brought together in convergence.* This explains the laser.

- The Kremlin, in a desperate attempt to salvage the Soviet Union's economy, set the stage for a few of its satellite states to restructure their ties to Moscow. This maneuver backfired when numerous demands for even more freedoms *converged externally* into a singular cry that swept across all borders like a tidal wave, consuming governments and nearly destroying the doctrine of communism in its wake. This explains the fall of the Berlin Wall.

- Millions of diverse and different peoples throughout the world have experienced a colloidal shift in consciousness (a brain shift) within the last two decades; their numbers increase hourly. Aftereffects from this powerful *internal conversion* indicate that structural and chemical changes are occurring in the brains of those involved, changes indicative of higher brain development. This explains the growing emergence of *the higher mind*!

Notice the timing of these examples—*right* now—on the edge of the twenty-first century, just as humanity faces the overwhelming challenge of establishing a global economic community and regional trading blocs.

So, how do we protect ourselves once available information exceeds our ability to process it? How do we even survive the avalanche of power unleashed when both personal and societal consciousness shifts and expands simultaneously? How do we handle the chaos that precedes new order?

Consider these voices:

- James Lovelock claims the interconnectedness of life systems proves that *cooperation, not competition*, best describes evolution and the most effective way to live.

- Physicist Ed Wagner says the *intercommunication possible between species* can be cultivated for mutual benefit. He points out, for instance, that trees employ a specific language based on wave frequencies he calls W-waves. Since they can, using this frequency, cry if alarmed, and "speak" to each other, he proposes that we verbally speak to them, too (that is, tell a tree you want to chop it down and explain why before you do; chances are, if your reason is a valid one, the tree will release the firm grip of its roots when the time comes, thus making it easier to remove).

- Michael Crichton, M.D., author of many books, including the novel, *The Andromeda Strain,* says *healthy thinking* will soon become not only an integral aspect of medical treatment but a scientific branch of medicine dealing with the mind's role in healing.

- Alvin Toffler, in his book, *Powershift,*[57] says *freedom of expression* is a precondition for economic viability in the future, as is *decentralization*. What is now emerging globally is no longer the need for mass political ideologies, but, he believes, highly charged and fast-moving "mosaic democracies" that correspond to the rise in diversity and operate according to their own rules. These new systems will force us to redefine the most fundamental democratic assumptions, as regions and localities become less uniform.

Self-healing always leads to self-governance and opportunities for higher brain development.

As people heal, as countries heal, political realities change accordingly. Contrary to popular opinion, morality improves with the expansion of cultural diversity, and efficiency and creativity advance with decentralization.

It has been said that a potentiality that never becomes an actuality is of no importance. Hence, a shift in consciousness is not enough. Value and meaning are determined by aftereffects—how we demonstrate what we think we know and the extent to which we share what we have learned with others.

Any form of growth is cyclic, not just from life to life but from one vibratory phase to another, in and through dimensions without number. Yet, only consciousness progresses, and the only true goal is to express and eventually rejoin the thought which stirred (our Source). We can be taught all manner and types of knowledge, but truth, the real truth of life, cannot be taught; it is "known" or "remembered" through the higher mind (soul mind).

We say that "coincidence is God's way of remaining anonymous" instead of admitting that the average person seldom recognizes preexistent information, or its ripples of give and take, for what it is. But the collective (that accumulative field effect of group consciousness) *always "signals" back to us what it recognizes from what we project into it, through the medium of symbols, trends, images, and through creativity and invention.* Examples: the movie *China Syndrome* foreshadowing Three Mile Island, the novel *Futility* foreshadowing the sinking of the Titanic (both mentioned in chapter 2), the resurgence in the popularity of Depression glass and other styles from the late 1920s just before the stock market nosedived in the late 1980s, and so forth.

(Along this same line, it has been established that the closer to the time a major incident is about to happen, the more precognitive "messages" there are from the public "announcing" the event.)

The purpose of these "signals" from the collective is to catch our attention so we can be aware in advance of the direction the group mind is "traveling" and what might result.

Life responds to itself. When we "remember," we know.

I submit the following as a descriptive explanation of the way group mind (the collective) responds to individual consciousness:

Learn a new word and suddenly your world will fill with that word or examples of it. Study a new subject and watch your life cram itself with what you are studying or opportunities to learn more about what you are studying. Think about something for a while and not only will it manifest, but wherever you turn, something akin to your thought will await you. Make a decision to move, to leave where you are (whether verbally or only in consciousness), then stand back. That back step will enable you to prepare for the rush of opportunities that will tumble in your direction—opportunities to keep you where you are or to test your resolve or to see if you really meant what you decided.

Life interacts with its participants (the process of the collective mind interfacing with individual mind).

It's as if the very air we breathe is alive, observant, and listening. Our thoughts, our feelings, our desires, our questions, our hangups, all are somehow heard or sensed, then acted upon—one way or another—to our benefit or to our detriment (sometimes called synchronicity). Belief or disbelief in this process seems to make no difference whatsoever. Life always responds (through the pulsating web-matrix), and we are part of this thing we call life because we are here, embodied, and breathing.

And it is a process, this thing that responds to us, a process of inter-connected pulsations interacting within the network that connects them (how the web-matrix works). Life supplies us with what we most need to develop our potential because we "tell" it through our responses to its responses (accessing the web-matrix). Like give and take, the process is elastic and flexible and more than accommodating. But the process can be a merry-go-round of survival instincts and foregone conclusions until we awaken from the nearly hypnotic spell it weaves, and realize we can choose differently; we can rise above and go beyond what seems to limit—we can step off the merry-go-round (which is the pressure from society to view everything in accordance with set standards and dogmas). When we do, alternatives shift and options change. Again, life responds to us, only this time results are different because we are different. We graduate from coex-istence to coparticipation when we consciously and decisively take charge of ourselves and the life we lead, and choose anew (when we see through "conventional wisdom" and recognize creative potentiality).

Jiggle the ether and things happen. But when you know what you are doing when you jiggle the ether, you can "move mountains." Why? Because the ether powers The Void, that ultimate womb from whence comes the substance of manifestation. And it is the ether that gives The Void its shim-mer, for the ether embraces the presence of indivisible consciousness every-where present (termed The Holy Spirit, or the numinous).

The "knowing" we can achieve is not based on education. The word "education," as you will recall, originally meant "to draw from that which already was known." In times past, people realized that intelligence was an extension of memory and memory fields (mass mind), and related to a per-son's ability to access preexistent sources of information. Even though this original concept has altered radically over the centuries, remembering the already known is still considered an expression of the highest level of knowl-edge a person can attain (sometimes referred to as mysticism or gnosis).

In modern times, many people call the ability to access higher knowl-edge or knowing "intuition." As defined in the dictionary, intuition is direct perception of truth independent of any reasoning process, clear and quick insight without prior knowledge, direct cognition. Others label this ability "psychic," which is defined as "of the soul," supernatural, outside of physiological processes or scientific knowledge, sensitive to subtle influences. Rather than using either term, I think it is enough to simply realize that the ability to avail ourselves of higher knowledge or knowing happens automatically after we alter or shift our consciousness and access other frequencies of vibration (other levels of memory and preexistent information).

Don't forget, children mentally flow in and out of such states all the time. Their secret magic? Receptivity.

Actually, we could learn a thing or two about receptivity from the work of Sir Issac Newton and his professed belief in magic.

Newton is credited with writing hundreds of thousands of words on various aspects of the subject. His huge personal library included 170 volumes on magic and over 300 on alchemy. This is why Newton added the colors orange and indigo to the five seen through a prism. By adding them, the color total corresponded with the mystical number seven, in keeping with his occult views. Historically, Newton is credited with "inventing" calculus (he actually refined what was already in use), the meticulous measurement of observable details, and the duplication of research results. His methodologies established the Newtonian Fundamentals, which still undergird the applied sciences today. Mostly forgotten, though, is the fact that Newton's initial goal was to take magical traditions and make them practical in the everyday world. He discovered the real meaning of magic in doing this, with humankind the benefactor.

You see, the real meaning of magic is "receptive." That's because the concept of magic originated from traditions of the Babylonian and Persian word for "receptive," which was then written as "magno" ("magnet" and "magnetic" derived from that term; so did "magi" and "magic"). These ancient peoples understood that when someone was receptive, or displayed receptivity (a willingness to receive), that person could then draw to him or her all manner of unique or desirable happenings with little or no effort, almost as if "charmed" (possessed of magic). This understanding of magic, of receptivity, opens wide the door to altered brain states and colloidal shifts in consciousness.

Actually, there are three ways to experience information:

Normalcy: What is perceived within the range of our physical embodiment, then processed through our internal sensory and emotional systems and clarified through reasoning and logic.

Reality: What results when our perceptual information reflects back to us from the ether matrix (the mirror effect), thus creating a backdrop (like adjustable scenes on a stage) that changes as we change, enabling us to create our own reality by the way we mirror our perceptions through the external environment of the life we lead.

Truth: What preexists as a central core of all intelligence; infinitely reliable, infinitely consistent, infinitely aware of its own nature as All That Is, and infinitely available to anyone through the higher mind (soul mind).

> We do not "arrive," we process.
> We do not "learn," we remember.
> We do not "become," we are.

I find it fascinating that in zoology it has been established that people when blindfolded instinctively travel in spirals, not circles, that the spiral movement is the universal property of living matter in motion. Since the spiral is how the thought which stirred forms itself and appears to move, we are imprinted by that mark, the spiral. Everything is. Life's expression only seems to be circular. That's because our eyesight is mostly forward focused, no matter how we turn our head, rather than off-angled and sideways, where the spiral can be seen. This restriction in our perceptual field is how we miss the torus and see either light waves or particles instead of both simultaneously. If we could train ourselves to develop wide-angled, peripheral vision and a greater sensitivity to subtle interplays of form (seeing sideways)—*and we can*—we could then catch the underlying truths we generally miss.

We enter this world on an in-breath.
We exit this world on an out-breath.
Back and forth; motion and rest.

20

Shadows and the Third Way

*Where love rules there is no will to power. And where power predomi-
nates there love is lacking. The one is the shadow of the other.*
—Carl Gustav Jung

Good and evil . . . a formidable topic. Visionary truth and our journey through time, space, matter, and mind mean little if we cannot
address the dynamics of good and evil and ask this question: How does
one cope with the shadows Light brings?

Since everyone loves a good story, I'll begin with some stories. Only
this time, you have a job to do, and that job is to identify the "victim" in
each vignette, all of them true, by the way, and from my own life:

- My husband and my son went elk hunting in the high country of
  Idaho with a friend. There wasn't enough money to buy all the grocer-
  ies we needed, much less meat. The trio had to bag an elk. They had
  to. It would be a lean and hungry winter if they didn't. My son silently
  pointed to a nearby ridge. The friend saw the huge bull elk he was
  pointing to and shot. Missed. Instead of running to safety, the elk ran
  toward my husband and suddenly stopped, barely two hundred feet
  away. My husband took his turn. Missed. He shot again. Missed. As
  long as my husband had his gun raised, the elk never moved, not one
  muscle twitched. He just stood there. As my husband later told it, the
  moment he readied for the third and fatal shot the elk calmly looked

him straight in the eye. As the elk did this, my husband felt a strange, weightless, floating sensation come over him as if he and the elk were merging. And he distinctly heard the elk somehow say to him, "It's all right. Take all the time you need. I will be the meat your family needs this winter." After the elk fell, the three men were beside themselves with excitement, as none had ever seen anything like it before. Our share of the kill was more than enough to last the winter, but the meat tasted different. It was sweeter, more tender and delicious than any meat we had ever tasted. We came to regard the meat as a sacred gift, and eating it became a communion with Love.

• Being the wife of a crop duster pilot and the daughter of a police officer, I have had many opportunities to be on the "front line" of countless accidents. Many of our close friends died in blazing infernos. Just before midnight a number of years ago, two friends, each in a separate plane, were spraying fields near Adrian, Oregon. They met head-on, sending a shower of debris earthward. Several large pieces of one plane exploded into the farmhouse below. Both pilots were killed, as was the woman in the farm house. It was a nightmare and a tragedy. Yet I suspect they all knew in advance they would soon die. The reason for my suspicion is that each of the three, as we later learned, had exhibited the same behavior pattern before the crash. They had each been compulsive about winding up their personal affairs, visiting with all their friends and loved ones and speaking deeply and intensely to each. And, they had checked and rechecked their insurance coverage and debts payable, making certain their spouse knew and understood how to manage the family finances. When this was done, they had each completely relaxed and with a peculiar glow about them, as if they were ready for something special to happen and all was right with the world. They all died forty-eight hours afterward. I've run into this behavior pattern before. It is characteristic of every pilot I have ever known who died in an "accident," and it is true of many other people I have had contact with who later died "suddenly." Because I have encountered this pattern so often, I have come to suspect that all of us, at least subconsciously, know when we are about to die; and, to the best of our ability, we prepare ourselves and those we love for the approaching finality.

- One of our daughters was abducted from the city park by a man with rape on his mind. A contingent of police officers and I were able to track down the culprit and rescue our young daughter, scarcely in time. When the case went to trial, the judge gave me an opportunity to help him decide punishment. I expressed the fact that I thought the man wanted to be caught, since he had left a trail that could be followed and his actions were consistent with someone who was "begging for attention." Also, because of all she went through, my daughter had been jolted from years of ambivalence about school to wide-awake alertness. Her grades jumping from Cs and Ds to As and Bs. Although I could not and would not condone what had happened to my daughter, it was as if the man had done her "a favor," and I felt it was now time to do one for him. I recommended counseling and vocational training instead of a prison sentence. The judge agreed and such was given.

- He was headed right for me and I was alone. He stood between me and the bright lights of a nearby hotel. Behind me was the dark span of ocean. It was late and the highway of beach was vacant except for the two of us. I began walking at a normal pace back to the hotel, but out of the corner of my eye, I noticed his stride quicken in my direction. He was muscular, like a dock worker, and he was wearing tight jeans and a T-shirt. Instantly, our movements froze while time and space expanded, sending a shower of sparkles everywhere. My possible attacker and I merged into one. His thoughts were mine, mine were his. Yes, he wanted my purse. The money. The eyes of my mind quickly scanned every item in my purse, searching for anything of value, wondering what decision to give him. No, the money would be of no great loss. Yes, I could replace everything, although it would be troublesome—but wait, my children's photographs—they could never be replaced. "No," I said to him, "I will not cooperate. What are my options?" "Well," he mumbled, "I just ate a heavy meal and I want an easy hit. I won't follow if you run." Just as suddenly as they had expanded, time and space contracted back to normal dimensions, and animated behavior resumed. I ran! Like a track star! I made it to lights and people. A brief glance over my shoulder revealed that the man had tried to catch me but missed. He was too slow.

Did you locate any victims in these stories? If everything is relative to the perception of the perceiver, and past/present/future simultaneous, how can there be victims?

I instinctively labeled my three near-death experiences of 1977 as "The Heavenly Sledgehammer Effect." Regardless of appearances to the contrary, I recognized that I was truly being given an opportunity to turn my life around in a major way. Never did the question "Why me?" seem valid; rather, "What can I do about this?" felt the better response.

My initial encounter with death happened because of a miscarriage and extreme hemorrhaging. I had been raped in the sense that I had never given the man involved permission to have sex with me. A car accident had forced him to seek temporary shelter, and since I knew him, I had volunteered the extra bedroom in my home for his use. That night, while I was sleeping soundly, he had crept into my bedroom and my bed. I didn't awaken fast enough to prevent what came next.

Does this make me a victim? According to the law, it does.

The first physician I went to didn't have me hospitalized, although his nurse was vocal in saying I should be. He administered an injection to stop the bleeding—an injection that he knew by reading my files I had no tolerance for—after he ignored my pleas about the inordinate leg pain I was suffering. Each time I asked him about my legs, he responded by laughing at my predicament and the fool he thought I was for ever letting the man in my house to begin with. Any defense on my part he rejected. The large blood clot in one leg, which set the stage for my second near-death episode, formed six inches below the injection he gave me. (Two other patients of his sued the man for malpractice several months later. I chose not to join them, as blood money did not interest me. I only wanted the public warned about the doctor. They were.)

Does this make me a victim? According to the law, it does.

People are victimized every day, especially children. I do not deny this. In fact, about 90 percent of the world's population seem to have little or no control over their lives. We all, myself included, must be alert to this injustice and vigorously work to prevent as much of it as we can whenever and wherever possible. Such activism has punctuated my entire life; I hope it has yours, too.

Although ready solutions to life's many shadows still elude even the wise, my experiences in life have taught me to view what appears as "evil" differently. This is because, invariably, just below the surface of what appeared to be true, I would encounter an undergirding reality that didn't match. In other words, I have learned that appearances can be deceptive—that good and evil, shadows and light, *are effects, not causes,* and effects can be changed. And to change effects, alter how you respond to them.

Stay with me on this.

All of us have a tendency to ascribe to people, places, events, and circumstances, power and postures not really there. Admit it, this strengthens effects (consequences). Hence, we can and often do create our own devils and demons, dungeons, and disasters by the way we focus our own mind—not necessarily because we've perceived a valid truth, but because we have decided we were "right" or were educated into thinking we were.

This tendency we have to allow society and others close to us mold and manipulate our perception of reality can be reversed, and effectively, once we learn how to discern the difference between cause and effect. We set this process in motion by owning up to and being responsible for our attitudes and opinions, our thoughts and feelings. We each have a tremendous latitude to effect change when we accomplish this feat, even if faced with impossible situations.

A perfect illustration of this comes from a newspaper clipping about the stirring message delivered by Rev. Jesse Jackson to the inmates at Chicago's Cook County Jail on Christmas Day, 1994. If you missed reading the story in your local newspaper, here is a brief excerpt:

> You have the power to change violence and crime just by changing your mind. You have the power to change the gun market, the drug market, the structure of America. You have the power to save our children.
>
> There is a 75 percent recidivism rate in Cook County. If you can cut that back to 50 percent, you would change the criminal justice system. Cut it back to 30 percent, and you would just about put the prison system out of business. And

you can do it if you change from the victim complex of self-pity and accept a new identity. The key to change is in your mind, in your heart.

And when we do this, accomplish the so-called impossible, we claim, "It's a miracle," never admitting that we were coparticipants in the drama that unfolded. To whatever degree, we have a part to play.

Incidentally, "miracle" comes from the Latin word *mirus* which means "wonderful." And that's exactly what it is, wonderful, when we can see through the illusion of our own perceptual preferences and discover the vast amount of power and control we actually have in our lives. The word "illusion" has another meaning, too. From its Latin root *illudere*, it translates as "inner play." You guessed it, on the deepest level of truth, *life is a game we play*. In the late 1800s, Florence Scovel Shinn wrote a little book that has become a perennial best-seller. *The Game of Life and How to Play It* explains the enigma of how to turn obstacles into advantages. More than a collection of principles, it is a book of action as valid today as when written.[58]

Think about this:

We have horror in life because hate is easy. There are bad guys and there are good guys because so few people are willing to face all aspects of themselves and dedicate the effort it takes to change the only person they can change—their own self.

We have horror in life because the thrill it gives us confirms that we are alive. We like to be terrified or angered or upset because it gets the adrenaline flowing and heightens our faculties and enables us to feel the body we live in and the life we lead. Sex doesn't last, but anger can, and so can fear. So, we seek them, in the name of boredom or power or excitement or defense or curiosity or for a million other reasons we seek them—and whatever is dwelt upon long enough happens—we manifest or attract to us whatever we seek.

We have horror in life because it hasn't occurred to most of us yet that maybe, on some level of reality and in some manner, we are either its cause or we have added to what caused it by our participation. Whether ignorance or manipulation or indifference or desire or a need to be right was the motivation, the resulting situation will not change until its participants do. And that means us, for you cannot have a war if no one shows up to fight.

Stop for a moment.

Compare the crowds who turn out to see a *Friday the Thirteenth* movie with those who go to see a film like *Chariots of Fire* or *A River Runs Through It*. Most people prefer the adrenaline rush of fright over that of inspiration and upliftment. Why? Because fright requires no particular commitment from us. The fear is free.

I was talking with a newsman at the radio station where my husband Terry once worked, and I asked him what types of events he most liked to cover. "No doubt about it," he replied, "murders. They're so fascinating. They make me feel alive and there's a lot to them. They're juicy." (Case in point—the O. J. Simpson trial. Did you follow every gory detail, or did you flip channels and watch something else on your television set?)

Nothing really changes until we do. We must be the change we want to see happen *before* it can. For instance, if we long to find someone we can trust, we must first be trustworthy ourselves. If we yearn to be loved or to be heard, we must first love others and listen to what others have to say. To have a friend, be a friend. To receive, give and give abundantly.

We want to know before we will believe, yet we have to believe before we can know.

Receptivity is the first step, then faith, then action.

But to recognize anything, just to see it, is to participate in its existence and give it the power to continue. Psychic premonitions exempt no one from this fact, nor do visions—for vision is simply the ability to see beyond the view. This means we are responsible for our reactions to whatever we recognize. As a truism, *wherever we put our attention is where we put our power*. And that power of ours gives power to whatever we recognize.

Read the following quote carefully. It comes from the address given to the people of Czechoslovakia by their appointed president, Vaclav Havel, on New Year's Day, 1990. It appears here with Havel's permission:

When I talk about a decayed moral environment . . . I mean all of us, because all of us have become accustomed to the totalitarian system, accepted it as an inalterable fact and thereby kept it running. In other words, all of us are responsible, each to a different degree, for keeping the totalitarian machine running. None of us is merely a victim of it, because all of us helped to create it together.

Havel is both perceptive and courageous. He saw beyond appearances (effects) to the real cause of his country's dilemma, and then reported what he found. He recognized this truth: Evil exists in the world because it is allowed to exist by the very people who ignore or refuse to accept their responsibility as coparticipants in whatever drama consumes them.

Take AIDS.

Paul Ewald, an evolutionary biologist at Amherst College, discovered that HIV (which causes AIDS) has been a benign and little-known contagion for centuries. The virus did not become a threat to society until after the social upheavals of the sixties (even considering the theories that it was spawned by a genetically altered batch of polio vaccine given to Africans in the late fifties or created to annihilate homosexuals throughout the United States in the early seventies). Add to this the comment made by French medical historian Mirko Grmek when he noted that, although Western-style monogamy is uncommon in Africa, mass migrations from the African countryside into overcrowded cities shattered the social standards that had once held sexual behavior in check. Consider, too, that the only compound yet found to be nearly 100 percent effective in treating the disease (Calanolide A, from the sap of gum trees in the Malaysian rain forest) was destroyed utterly before it could be synthesized—by locals hired to clear trees for land.[59]

No matter how you examine the AIDS tragedy, this observation arises: Both the spread of AIDS and the destruction of Calanolide A were the result of behavior changes from people seeking short-term answers to long-term challenges.

It is said that we are not punished so much for our "sins" (mistakes), as by them.

For the most part, we believe that we are "victims" because we are conditioned by the bias of "conventional wisdom" to think we are. We don't question. We don't do our own thinking. Remember the replies made by so many Germans after World War II ended, when they were asked about the concentration camps and the holocaust and why they did nothing to interfere? "That's not my concern." "That was none of my business." "You don't question authority." "I always do what I am told and I was told to be quiet." "What my government leaders do is always right." "One person doesn't make any difference."

Thousands of years ago, the elder rune sign for "poisoned dagger" (part of an ancient Russo-European form of ritual hieroglyphs, which is described in two books of mine, *The Magical Language of Runes* and *Goddess Runes*[60]) included in its design the hilt of a dagger as well as the blade, implying collusion (hilt/hand/decision; blade/problem/result). These early peoples understood that there was a connection between any given problem and the one who had the problem—whether that connection was a word carelessly spoken or conveniently omitted, a decision that put others at needless risk, or activities engaged in without regard to consequences.

Without hesitation I can say that the publication of *Coming Back to Life* was the single most negative event I have ever lived through. Not only were my husband and I nearly bankrupted by the publishing debacle and a dishonest agent who stole over two-thirds of my advance money, but also certain persons in the field of near-death studies were spiteful and vicious in their unwarranted attacks against me and the research I did. This put the book at further risk and resulted in "unofficial" censure. I became a cynic because of it. This state of affairs healed once I realized that my cynicism was nothing more than a smokescreen I had erected to mask my own disappointment with myself and my behavior.

Shadows, including the hypocrisy I let overwhelm me, represent denied aspects of the self; they are that part of our own being we do not want to recognize, face, or admit. Since there is no way we can hide from the self we are, anything denied will be reflected back to us through circumstance—either from another person, a physical event, or because of some activity. Life is a mirror, and it continuously reflects whatever imagery is presented to it.

(The creative medium of the ether returns to us what we project into it like a boomerang. In my case, I finally recognized that I was afraid of success [that is, public accolades terrified me]. This fear of mine bounced back to me through life's mirror in the guise of personal and professional attacks and censure. The more I complained about this perceived "injustice," the more I subconsciously attracted to me one "opponent" after another, thus enabling me to avoid facing my own shortcomings by busily pointing out everyone else's.)

When we are faced with "reflection from our mirror" (the physical manifestation of our projected attitudes and opinions, thoughts and feelings), we tend to react in one of three ways: (1) we play ostrich and pretend it away; (2) we label it an enemy or a devil and attack; or (3) we confront the issue squarely and honestly, search for the truth behind the appearance, and take decisive steps to initiate a constructive solution. The first way creates victims, the second victors (conquerors), and the third responsive and responsible participants in life committed to growth and learning.

This third way of dealing with life issues is the way in between duality—in between victors and victims, good and evil, shadows and light. The Third Way requires mediation and diplomacy skills, mindful attention, plus a willingness to consider *appropriateness* as a greater priority than self-centered interests. It takes time to learn and patience, and it necessitates cooperation and compromise, but it is the only way of living that shows any promise for the kind of future that affirms the worth of life. The Third Way upholds dignity and value and wholeness—and *wholeness is spirituality made manifest.*

We move past duality with The Third Way and not by avoiding the challenge opposition brings, for a certain amount of conflict ensures growth. (Seeds only germinate in darkness, you know.) By recognizing effects to be what they are, illusions (the inner play of possibilities), we are empowered to choose the focus that is the most appropriate for our highest good and the highest good of all concerned. When we are mindful of the real truth of life—the divinity of wholeness—we are led to positive solutions that are health giving.

The Third Way leads to the path of radiance, the path beyond opposites and role playing, into the dimension of patience, where the center-point resides.

When we live in accordance with The Third Way, there is less tension.

Karma is the law of cause and effect (that is, what we sow we reap), and it is based on tension. Yes, without some degree of tension, neither cause nor effect could exist. Still, the best way to reduce karma is to forgive, because *forgiveness dissolves tension and promotes patience.* This is why we can never transcend what we resist. We need to let go, to grow. We need to forgive.

The first two dimensions of existence identified by Edgar Cayce are based on tension, the tension necessary for time and space and matter to exist. The third dimension maintains the integrity of the other two. Yet, without all three, there could be no creation, no earthplane, no us.

When we slip between the cracks of perception, of what appears as solid and real, we encounter what is referred to as "Truth." Truth is preexistent information with enough power to retain its original form.

Since the only real geography is consciousness, it is not where we are but what we have become that makes the difference.

We do not exist in the earthplane so much to transform ourselves as to experience ourselves. Once we have discovered our true nature and our true worth, once we have slipped between the cracks of our own perceptual preferences, *we automatically transform* from the experience . . . we become who we really are.

As far as I am concerned, the only purpose of shadows is to provide us with enough contrast so we can awaken to and recognize our true identity, and so enough fast-slow rhythms can be produced (change/integration) to guarantee that we evolve.

Love defines this awakening, for love is the force which stirs the Great Thought and infuses us with its divinity. This love, God's love, is our birthright *and* our salvation.

An old adage says, "God can do no more for you than through you." This is because preexistent information rushes to our aid via the webbing only when invited. We must take that first step. We do this by inviting God into our lives through prayer or by simple request, and then actively listening to the fullness and the power of God's silent reply.

The Third Way is expressed in nature as The Golden Mean.

This mathematical formula celebrates the unique relationship between two unequal parts of a whole, where the small part stands in the same proportion to the large part as the large part stands to the whole.

Experiments conducted thus far to record and measure wavelengths from the emotion of love have all produced the same configuration—The Golden Mean.

Pattern of the Golden Mean, found in the nautilus shell.

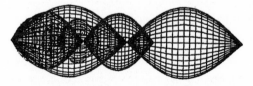

Wavelengths of the Golden Mean, found in the energy of love.

The equations below were inspired from the book *The White Hole in Time* by Peter Russell.

Gravity = the attraction of mass for itself

Love = the attraction of life for itself.

21

Centerpoint

If the only prayer you say in your whole life is 'Thank You,' that would suffice.
—Meister Eckhart

Physicist Wolfgang Pauli once decreed that a new science is needed to explore the objective side of human consciousness and the subjective side of matter. Not mysticism, but a science willing to incorporate objective and subjective avenues to discovery while recognizing the legitimacy of personal experience. It is my hope that this book is a step in that direction.

People who rely on only one or two modalities of information gathering miss more than they gain, be those individuals psychics or scientists.

Perhaps this truism explains the urge that drove me to compare what had happened to me at the edge of death with the experiences of thousands of other people who had also undergone the near-death phenomenon—then question, probe, test, and double-check what I observed. The first time I launched a project of this scope was in the midfifties, after an illumination of the Holy Spirit had bathed me in a singular ray of light as I stood in church during worship services to commit myself to God. The second was ten years later, when a kundalini breakthrough left me unable to eat properly or respond to normal stimuli for a period of forty-four days. Reading Bible passages, esoteric texts, and legends of the sacred mysteries was never enough to satisfy my curiosity. It was research that made the difference. Actually, I began doing double-blind experiments at the age of five.

My many probes into the meaning and the purpose of existence have convinced me that we humans strive to see life as we want to see it, rather than as it is. We comfort ourselves with whatever bias we prefer, especially if that bias is shared and supported by others. Some people even risk life and limb to prevent the dissemination of any information that might challenge traditional or treasured beliefs. I find this behavior odd, for if people would simply relax into the silence of their own soul, they would uncover marvels beyond mind or measure. Even just being able to glimpse a broader view of "the big picture" can free one to reevaluate whatever was once taken for granted and initiate positive and constructive lifestyle adjustments.

As more and more people do this, alter and expand their consciousness, the stories they tell about what they encounter are helping all of us to see ourselves and our world differently. No one, though, no matter how much was revealed during the otherworldly episode, comes away from his or her experience with as much knowledge as it may seem, me included. It is the sum of everyone's stories that matters most, not just individual renderings. If you use near-death studies as a example, you will notice that the idea of "amazing grace," of individuals being singled out by God, is not nearly as persuasive as the overall impact of all the stories in combination. Then you see the pattern, the pattern of an emerging global consciousness that is of staggering proportions. You also see, and quite clearly, that the pivotal experience of transformation is but one of many steps along an unending spiral of growth and learning—not a means to an end.

Thus, a transformation of consciousness is seen for what it is—a healing into wholeness, a nudge toward higher octaves of brain function, personal integrity, and creative problem solving. People who go through a brain shift awaken to the power of their inner self and to the strength of community, not as an excuse or an escape but as a way to make new.

This higher octave of expression is why the Perelandra Gardens described in chapter 2 is so successful. The concept of "energy gardening" that arises from Perelandra recognizes that you do not plant seeds and bulbs as separate items or in random rows in a garden plot. Rather, you plant members of a whole *in ratio to the energy of the whole*. When you keep plants in balance with each other plus the environment, the energy configuration that results maintains its own health—regardless of threatening diseases or pests or unexpected changes in the weather. (This is the principle of The Golden Mean.)

The same energy configuration shows up in matters of bodily health. For instance, if a woman needs more calcium, you do not just give her more calcium. You balance her nutrient levels in ratio to the needs of her *entire* body-mind complex—and you do this by lowering protein intake (which uses up calcium), increasing carbohydrates (which enhances calcium levels), while suggesting that she explore any mental or emotional concern she may feel about the support structures in her life (calcium relates to bones, which relates to support structures). By doing this, the integrity of the whole is reached and balance results (the coherence of true health).

This principle applies as well to economics, manufacturing, and politics, and it is why communism failed. Communism's enforced status quo violated the very balance life insists upon. (One proponent of balancing the many parts within the whole and of creating higher ratios of integrity [coherence] is Dean Black, Ph.D. He uses the label "contextual healing" to identify the system of wellness that is based upon the body's native intelligence to heal itself once the needs of the larger whole have been addressed.[61])

I am reminded here of a new version to the Pledge of Allegiance that was penned by the well-known author and consciousness researcher, John White, and it goes like this: "I pledge allegiance to Humanity and the planet on which we live. One world, under God, indivisible, with peace and enlightenment for all."[62]

As our journey into the mind draws to a close, I would like to admit that I have not always handled well the many tasks I have chosen to accomplish in my life. I have learned, however, that by my participation, by the power I bring to what I do. I can embrace both darkness and light in celebration of the pathway that leads beyond polarity into the radiance of The Third Way. Waking up to greater truths means just that, but it also means taking responsibility for yourself and the way you live. In light of this, I cannot change how I was raised as a child in this lifetime nor what I have already done about the many opportunities presented to me. But I can change my understanding of and my response to the gift of life I hold in my heart—now and in the days to come. And the same is true for you.

With the challenges I faced in 1977 came joy unending, and for that I give thanks. Experiencing the centerpoint of creation/consciousness taught me a few things. Among them

- Always there is life. We cannot escape ourselves or what we have fashioned ourselves to be, as death ends nothing but the physical body we wear. The soul, who we are, continues.

- Each moment is precious and a moment well lived enriches all our forevers, and forever can be counted on.

- The purpose of history is not to limit our tomorrows but to free them, for as we learn the lessons our past would teach us, we are freed from the high cost of deception and deceit, freed to reinvent our world, not be imprisoned by it.

- The bottom line is not profit; it never was. The bottom line is service *plus* long-term investment in the education and the upliftment of others. The law of entropy only applies when greed or indifference underlies our motivation.

- Enlightenment is ongoing, not a plateau we achieve, as the term describes an evolutionary shift from one phase of brain function to another, opening the way for dimensions of experience without number and realms of spirit without end.

- The differing levels of heaven and hell are but stages of consciousness in The Grand Spiral of Remembrance. We reinhabit these thought-forms in an effort to cleanse ourselves as we prepare for the next octave of growth.

- There is only one disease, congestion of oxygen (energy), and only one cure, circulation of oxygen (energy).

- Illness has only one purpose, to deliver a message from the soul level to the personality level, for us or to someone else through us.

- There are only two religions on this earth, the religion of love and the religion of fear, and everyone belongs to one or the other, whether admittedly or not.

- The only gospel we can ever know is the experience of God in our own heart.

Love is the only standard.

Choice is the only process.

Forgiveness is the only protection anyone has, for you become whatever it is you cannot forgive.

God is.

God is love.

God as love is the only standard.

Truth can be summed in four words: One Mind, many thinkers.

Epiphany is a profound illumination of wholeness that occurs when least expected . . . to anyone. Truly, heaven is but a breath away. You do not have to die to find it.

Brenda Donaldson and Cindy Kidwell, both twelve years old at the time, were playing around with thoughts of their future. They made a discovery they said I could share with you:

Be what you is
because if you be what you ain't
then you ain't what you is.

The electric wave nature of the human heart.

I have sought to reweave
the fabric of your mind with this book.
Thanks to experimentation with positron emission
tomography (PET),
we now know the brain can rearrange itself
in as little as fifteen minutes—if excited
in novel and creative ways. Because the design of Future
Memory is based on a labyrinth,
all the "triggers" are here,
all the stages of the colloidal condition
patterned into its pages. Thus, it is possible
to simulate a brain shift . . .
just by reading this book.
Time is accelerating.
The atomic clocks in Boulder, Colorado,
made to keep perfect time without any influence save the
atomic energy that powers them,
have had to be reset
nineteen times since 1972.
Recent shifts in the earth's magnetic and electric fields are
affecting our immune systems
and fatigue levels,
not to mention our weather
and political-economic climate.
To keep up with our changing times, we too must change . . .
The future is now!
Return to an awareness of The Spiral's Edge
where all things can be seen.
Guidance from your soul awaits.

Appendixes

Appendix I

*There is a soul force in the universe, which if we permit it will flow
through us and produce miraculous results.*
—Mahatma Gandhi

There is a difference between *altering* one's consciousness and changing
one's consciousness, and that difference is important. Alterations of consciousness are always short term and usually result in simple, curious, or
unstable aftereffects. A change in consciousness, however, is always long
term, with significant or life-changing consequences a distinct possibility (that is, a brain shift). Since brain shifts are discussed at length in this
book, appendix I is devoted to a brief look at alterations of consciousness—by either artificial or natural means—and the effect these can have
on individual experiencers.

When consciousness is *artificially altered*, whether from drugs,
sounds, pulsations, incense, numbing sensations, or similar methods, the
new focus usually expands out almost immediately and is totally experienced as complete reality. Then, it further intensifies before ending,
usually as quickly as it began. Although exciting, most often the "high"
rapidly wears off and little of value comes from the experience. But when
consciousness *alters naturally* or in concert with a sincere desire for
growth and learning, the new focus usually takes time adjusting before
it expands out and is fully experienced. Then, as the focus intensifies, life
issues invariably surface, while what was previously believed often shifts
significance. Seldom does the experience end without imparting some
degree of knowledge or wisdom that can have a constructive and positive
effect on the person's life.

Here is the list of questions I have put together for use when attempting to evaluate the possible worth of altered states of consciousness:

Questions to Ask about Alterations of Consciousness

Overall Message from the Episode: What effect does it have? Does it inspire the individual involved? Does it empower him or her? Is there guidance given? If so, is this guidance positive or negative, clear or confused, practical or contradictory? Does the overall message enhance or demean the living of life? What does the individual involved feel about what happened? How and in what manner does that person respond? Does the overall message honor the individual's free will? Does it also honor and respect the free will of others and their right to self-determination?

Aftereffects from the Episode: Do the aftereffects increase or decrease with the passage of time? Are the consequences life affirming? Is the individual's character enriched or compromised by what happened to him or her? Is the aftermath from the episode constructive or destructive? Does the effect this has inspire and assist, or frighten and threaten, other people? In what way does the individual's life alter because of the aftereffects? Do these differences foster a more responsible and creative lifestyle, or paranoia, or depression? Does the individual become more or less fearful because of what happened? Does he or she become more moral and trustworthy afterwards, or less so, or unaffected either way? Is there a difference in the individual's health? If so, in what manner has his or her health been affected? Does the experience authenticate humankind's deepest and most time-honored teachings about human nature and the value of human life, or does it somehow seem to put at risk or invalidate these truisms?

This list of questions is helpful in dealing with a variety of subjective experiences such as visions, dreams, disembodied voices, channeling, religious conversions, angelic or spirit visitations, UFO contacts, near-death episodes, otherworldly occurrences, and transformational shifts in an individual's consciousness and character no matter what the manner of that transformation. Of concern always is how uplifting and constructive the message and how empowering and enriching the aftereffects.

Any form of consciousness shift initially causes some degree of disorientation and confusion. But comparing artificially induced experiences with those that happen naturally is like comparing lollipops with vegetables. Both are food. Both are ingested. Yet, it is the natural version that is growth affirming, while the artificial variety is geared more to the pleasure of the moment.

Drugs mimic the colloidal state, that is what the high is, but they do not and cannot complete the colloidal process. They reinstate tension before the expansion can be stabilized. Unfortunately, the tension they reinstate is easily distorted, creating the desire or need for more and greater highs.

In native societies, drugs were used as part of a ritual—never alone. It was the process of that ritual, the intention and meaning behind it, along with the supportive interaction between participants and elders, that balanced any drug's effect. The drugs used today are dangerous. The risk is not necessarily from the experience as much as the aftereffects. Whether temporary or destructive, transforming or life enhancing, *always* there are aftereffects!

Appendix II

All of life is the prioritizing of time and space.
—Rev. Ben Osborne

Endnote 21 details how to obtain a complete copy of Jack Houck's paper, "Conceptual Model of Paranormal Phenomena." His work is distinguished, I believe, by his exceptional achievement with the PK Parties he has facilitated since 1981. (PK stands for psychokinesis, or "mind over matter.") As a systems engineer in the aerospace industry, his desire has always been to formulate an explanation of how this and other types of psychic phenomena may operate, and I, for one, applaud his efforts.

He notes: "The human brain is both a transmitter and a receiver of information . . . the mind is not local just to the human body. Information that is about events in all space and time is stored all around us. The mind accesses this information storage system. The brain processes information as a very advanced computer would from both our physical senses and from this stored information. The brain/mind can tune into any information in this storage system when given specific instructions about space and time. The more specific the instructions about the information desired, the better the quality of the received (or retrieved) information."

A particular diagram appearing in his conceptual model is relevant to our discussion of how past, present, and future might coexist. Thanks to his generosity, this diagram can be presented here, along with a brief explanation:

Jack Houck's Conceptual Model
of Space-Time Relationship

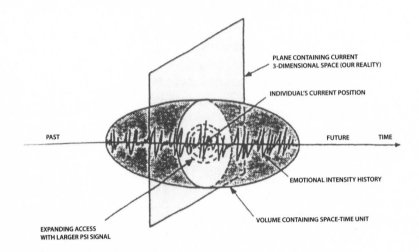

A shaded ellipsoid appears in the diagram; its shape is symbolic only, meant to represent the presence of all information about past, present, and possible futures. Designated as a space-time unit.

The square plane illustrates how our present three-dimensional reality intersects into the space-time unit where all information is available.

The center line is a time axis indicating past, present (located in the middle), and future. The pointed arrow indicates the continuous movement of time.

The white circle in the middle shows our present awareness of the physical universe at the current time. Note how this awareness is central to our being. Dotted lines circling out from the circle's core illustrate that as our awareness expands, so too does our access to information.

The wiggly zigzags along the center time axis are a record of emotional intensity and of when it peaks and falls. Everything in existence carries a record of its own emotional intensity; even inanimate objects like a mountain will register and store a peak "emotional" event such as a volcanic eruption.

Future Memory

This model suggests that the mind scans in time, locks onto a peak emotional event, and then is able to recall the information wanted. Since the human brain is unlimited by either space or time, needed information becomes accessible because of the way emotional intensity is stored in memory. Time shifts that occur during paranormal phenomena such as telepathy, psychometry, mind reading, past lives, and psychic healing can be studied in the context of the mind scanning forward and backward in time and then locking onto events that have emotional impact or emotional meaning.

Jack Houck found during his PK Parties that the key to the availability of memory and the success of individual experiences with paranormal phenomena is emotion. This means the limbic system must be stimulated, since emotions are a product of the limbic system.

Back in the sixties I discovered the same thing, that in order to obtain repeatable psychic phenomena, or even to teach about them, one had to raise an individual's emotions to higher, more intense levels. Once that was done, almost anything was possible. I find it interesting that both pagan rituals and charismatic church services are designed to help adherents achieve a fervor pitch of excitement so that phenomena can result—things like faith healing, fire walking, snake handling, speaking in tongues, prophesying, ecstasy, spirit possessions, and so forth.

In fact, almost any shamanic ritual, spiritualistic ceremony, or church service is based on the premise that you must *involve the emotions to spark interest, and excite the emotions to touch the soul.* Simply put, stimulate the limbic system.

During a near-death episode or in any condition of brain shift where a change in consciousness or enlightenment is possible, I believe the limbic is automatically activated or somehow altered and enhanced by the shift itself. Certainly, this would involve frontal brain lobes, the neocortex, as well as whatever secondary systems might exist; perhaps explaining why experiencers display so much psychic sensitivity afterwards.

Appendix III

We find great things are made of little things,
And little things so lessening till at last
Comes God behind them.
—Robert Browning

The charts presented here were the result of an unexpected phenomenon. While addressing an audience on one subject, my brain/mind assembly busily connected bits and pieces of research data on other topics into a specific charting system that had nothing to do with my talk. Suddenly, material that had perplexed me for years came together by itself and in precise detail. As soon as I could, I grabbed pen and paper and recorded the whole thing, refining later, as needed.

I was fully consciousness of living in and operating through two distinctly different dimensions when this occurred; yet there was no sense of any split, tear, or division in my reality. I want to make this clear. The phenomenon, and I call it "two for one," felt totally natural to me. It has happened on occasion since my near-death episodes; I have no idea how, nor can I explain it. I can only say that I wish it would happen more often, because it certainly saves me a lot of time and effort.

My thanks go to Oralee Stiles and Marzenda Stiles McComb, former co-owners of Stiles for Relaxation Bookstore in Portland, Oregon, for sponsoring the workshop I facilitated that day. Their support made a significant difference in creating the kind of atmosphere where the unexpected could seek expression . . . and flourish.

The charting system that came together in such a peculiar manner fills the rest of appendix III, together with my comments.

BASIC DEVELOPMENTS IN TIMEKEEPING		
WITHIN CONSCIOUS PERCEPTION		BEYOND CONSCIOUS PERCEPTION
EASTERN	WESTERN	GLOBAL
Water clocks Flexible hours Adjustable beats Patterns of time	Mechanical clocks Equal hours Regular beats Segments of time	Digital clocks Equal instants Computer bytes Nanoseconds of time (a billionth of a second)
Based on scoop measure from continuous water flow	Based on point measure from artificial oscillation	Based on electrical pulse stimulation of silicon chip circuitry
Accommodates changing calendars	Maintains constant calendar	Aligned to universal grids and computime
FLOW	PROCESS	INFORMATION

When you study the history of clocks, a fascinating correlation is revealed between the mindsets of Eastern and Western cultures and the global awareness so popular today. Notice on the timekeeping chart how the adjustable flow of water clocks was the preferred method of keeping time in the East, while those in the West insisted on the more precise oscillations of mechanical beats. Modern computers, however, are aligned to a billionth of a second in universal grids, thereby completely removing the passage of time from any dependence on nature or mechanics.

Lewis Mumford, the visionary philosopher and educator, spoke vigorously about how the new computime forces people into unnatural lifestyles. In several of his books, he made it clear that clocks were not invented originally to keep track of hours, but to synchronize the actions of the masses for the political and economic gain of those who held the most power. Before he died in 1990, he made a statement that is worth repeating here: "The test of maturity, for nations as well as individuals, is not the increase in power but the increase of self-understanding, self-control, self-direction, and self-transcendence. For in a mature society, man himself, and not his machines or his organizations, is the chief work of art."

Continuing with the same format of that which is within and beyond conscious perception, we can change the emphasis of the chart on time-keeping to one that examines the basic aspects of existence. In this second chart, notice the connections between metamorphous (a structural change of form) and the force of an implosion, between evolution (the process of developing form) and the force of an explosion; then note the unlimited potential available once there is convergence.

BASIC ASPECTS OF EXISTENCE		
WITHIN CONSCIOUS PERCEPTION		BEYOND CONSCIOUS PERCEPTION
IMPLOSION	EXPLOSION	CONVERGENCE
Black hole Singularity Chaos	White hole Simultaneity Order	Colloidal state Suspension In-Between
Collapses and compresses and pulls inward	Emerges and extends and pushes outward	Exists as a unique condition with its own separate laws
Negative Dark Waves	Positive Light Particles	Neutral Sparkles Void
Contraction Past Finish	Expansion Future Initiate	Rest Now Hold steady
METAMORPHOSIS	EVOLUTION	POTENTIAL

The next chart explores the limbic connection and both brain hemisphere functions. Pay special attention to how religion can be correlated with the subconscious aspect of mind (right-brain

THE MIND: ASPECTS AND CORRELATIONS		
WITHIN CONSCIOUS PERCEPTION		BEYOND CONSCIOUS PERCEPTION
RIGHT-BRAIN HEMISPHERE	LEFT-BRAIN HEMISPHERE	LIMBIC GATEWAY
Subconscious mode	Conscious mode	Superconscious mode
Endows and enhances	Separates and analyzes	Synergizes
Feels/Absorbs Creates Subliminal	Clarifies Plans/Schedules Over	Senses Embraces All-inclusive
Imagination/ Intuition	Intellect/Knowledge	Collective whole/ Memory fields
Explores unlimited possibilities as it collects ideas	Experiments with details to establish limits and values	Knows and knows that it knows; instantaneous gut response
Night	Day	Shift
Female qualities	Male qualities	Soul qualities
RELIGION	SCIENCE	GNOSIS

hemisphere), how science naturally develops from the conscious aspect of mind (left-brain hemisphere), and how the limbic system opens up mysterious passageways to the soul and to gnosis through the superconscious aspect of mind or the collective whole. Remember that the limbic system is considered the zero point, that area of shift where vibrational levels converge.

Now, by combining all the charts presented thus far, one large chart can be produced (see "Death – Life – Eternity" chart). Under the heading "Death – Life – Eternity," we can use the limbic gateway and right- and left-brain hemispheres to examine—via what is within conscious perception and what lies beyond conscious perception—the extraordinary links between time, space, and consciousness. This examination is possible because of the way each chart flows into the next. Creation, and how it could continue to exist, takes on new meaning here, enough to help us reconsider what we think we know about life and death and whatever else might exist within the eternality of Mind. Notice that this larger chart progresses through the same stages and factors and in the same sequence that most individuals do once their consciousness begins to alter. And, as discussed earlier in this book, one first loses a sense of time when this happens, followed by the dissolution of space and form. One message dominates this chart: All is Mind; Mind is All (see the "Death - Life - Eternity" chart on page 251).

If we consider the possibility of connections between the development of spirituality and religion and what we have already covered, then some interesting patterns emerge (see the "Religious and Philosophical Aspects and Correlations" chart on page 252).

Of the three readings (texts) mentioned in the chart entitled, "Religious and Philosophical Aspects and Correlations," I regard *A Course In Miracles* as more right brained in its approach, *The Science of Mind* as more left-brained, and the wonderfully simple *Love Project Principles* as a reflection of pure essence. Should you be unfamiliar with any of these, please refer to the resource section for details.

According to my way of thinking:

Religion seeks to standardize God.

Spirituality seeks to personalize God.

Mysticism seeks to allow God to be God.

Some truisms based on The Holy Trinity:

TRUISMS			
TYPE	TRIAD PRINCIPLES		
Perceptual Religious Scientific New thought	Individual The Son Matter I Am	Society The Holy Ghost Energy We Are	Collective Whole The Father Information God Is

The world's major political ideologies, at least conceptually, can also be compared to the three major sections of the brain. Doing this helps to illustrate how consciousness tends to arrange itself among world hemispheres as it does brain hemispheres:

Communism, socialism	Right-brain hemisphere
Capitalism, monarchies	Left-brain hemisphere
Democracy, the *I Ching*	Limbic gateway (centerpoint)

Yes, I paralleled Democracy with the *I Ching*. The *I Ching* or, *Book of Changes*, is one of the first attempts in recorded history to place humanity in a more universal framework. Over three thousand years old, the book is mostly thought of as an oracle (a fount of wisdom accessed through divinatory practices), when in fact it was originally written to be used as a political guide for helping people and nations govern in ways consistent with natural law.

Many other firsts can be credited to this monumental piece of work. Among them, the *I Ching* is the first major compilation devoted entirely to the principle of unbroken wholeness, and it is the first recorded work based on our "modern" two-digit binary system. Leibniz, the German mathematician and philosopher who is credited with discovering the binary principle in 1679, obtained a copy of the *I Ching* from a

DEATH – LIFE – ETERNITY		
WITHIN CONSCIOUS PERCEPTION		**BEYOND CONSCIOUS PERCEPTION**
RIGHT-BRAIN HEMISPHERE	**LEFT-BRAIN HEMISPHERE**	**LIMBIC GATEWAY**
TIME		
Water clocks	Mechanical clocks	Digital clocks
Flexible hours	Equal hours	Equal instants
Adjustable beats	Regular hours	Computer bytes
Patterns of time	Segments of time	Nanoseconds of time
Based on scoop measure from continuous water flow. Accommodates changing calendars.	Based on point measure from artificial oscillation. Maintains constant calendar.	Based on electrical pulse stimulation of silicon chip circuitry. Aligned to universal grids and computime.
FLOW	**PROCESS**	**INFORMATION**
SPACE		
Black hole	White Hole	Colloidal State
Singularity	Simultaneity	Suspension
Chaos	Order	In-Between
Collapses and compresses and pulls inward	Emerges and extends and pushes outward	Exists as a unique condition with its own separate laws
Negative	Positive	Neutral
Dark	Light	Sparkles
Waves	Particles	Void
Contraction	Expansion	Rest
Past	Future	Now
Finish	Initiate	Hold Steady
METAMORPHOSIS	**EVOLUTION**	**POTENTIAL**
CONSCIOUSNESS		
Subconscious mode	Conscious mode	Superconscious mode
Endows and enhances	Separates and analyzes	Synergizes
Feels/Absorbs	Clarifies	Senses
Creates	Plans/Schedules	Embraces
Subliminal	Overt	All-inclusive
Imagination/Intuition	Intellect/Knowledge	Collective whole/Memory fields
Explores unlimited possibilites as it collects ideas.	Experiments with details to establish limits and values.	Knows and knows that it knows; instantaneous gut response.
Night	Day	Shift
Female Qualities	Male Qualities	Soul Qualities
RELIGION	**SCIENCE**	**GNOSIS**
INWARD SPIRAL	**OUTWARD SPIRAL**	**BEYOND THE SPIRAL**
EXPRESSION	**PROJECTION**	**MIND**

RELIGIOUS AND PHILOSOPHICAL ASPECTS AND CORRELATIONS

WITHIN CONSCIOUS PERCEPTION		BEYOND CONSCIOUS PERCEPTION
RIGHT-BRAIN HEMISPHERE	**LEFT-BRAIN HEMISPHERE**	**LIMBIC GATEWAY**
Subconscious Aspect	Conscious Aspect	Superconscious Aspect
Imagination	Knowledge	Collective Whole
Religion—Intuitive Approach	Religion—Intellectual Approach	Spirituality—Accesses Memory Fields
Paganism	Church Theology	Mysticism
Psychic/Shaman	Metaphysician	Mystic/Saint
Phenomena	Tools/Techniques	Silence
Chaos	Order	In-Between
Endows and Enhances	Separates and Analyzes	Synergizes
Feels/Absorbs	Studies and Demonstrates	Senses
Personal Expression	Precepts and Principles	Gnosis
Visions and Dreams	Bibles and Scriptures	Faith
"A Course in Miracles" text	"The Science of Mind" text	"Love Project Principles" listing
Effect	Cause	Unification
Experimental	Interpretive	Grace
Subliminal	Overt	All-inclusive
Finish	Initiate	Hold Steady
Teachers	Leaders	Avatars/Messiahs
Female Qualities	Male Qualities	Soul Qualities
Horizontal/Curves	Vertical/Lines	Abstract/Zero Point
Waves	Particles	Void
Contraction	Expansion	Rest
Patterns	Symbols	Wholes
Inward Spiral	Outward Spiral	Beyond the Spiral
So-called "Evil"	So-called "Good"	Unconditional Love
Shadows	Sunshine	Radiance
Negative Light	Positive Light	God's Light
SPACE	**TIME**	**PATIENCE**

Jesuit missionary twenty years later and learned, much to his surprise, that what he had "discovered" was known about and used in ancient China thousands of years before he ever conceived the idea. With today's computers, the unbroken line of the *I Ching* system (the yang, which embodies the positive, male, assertiveness) is represented as a single stroke "I," and the broken line (the yin, embodying the negative, female, receptivity) is represented by "0."

Appendix IV

We are intellect without form. All else is choice.
—Greek temple inscription, circa 400 B.C.

Any discussion about the colloidal condition invariably leads to mention of "The Vertical."

Throughout all cultures and all ages and in all sacred teachings, there is but a single image said to signify transformation. Appearing variously as a pillar or column, hourglass, huge tree with roots and branches, ray or beam of light, upward path or road or stairway or ladder, pairs of cones or vortexes or cyclones (one inverted atop the other with spouts or points touching) . . . whatever shape taken, the resulting image is always vertical and it always dominates any given scene, be that of a dream, painting, or visionary experience. And always there is the sense that this large vertical object connects the earth with heaven, and that ascending or climbing up its length will lead one away from all horizontal concerns (the law of cause and effect, duality, and the stages and cycles of life) to the unlimited realms of grace and freedom and radiance that exist beyond ordinary reach. Traditionally, The Vertical refers to the act of redemption and resurrection and it symbolizes the spiral's edge—as well as the energy pattern of the number thirteen.

Since legend and lore about The Vertical abound cross-culturally and are timeless in scope, I'd like to share a brief summary of the main symbolism surrounding these teachings. Along with The Vertical, I will also address The Horizontal and its corresponding energy pattern in the number twelve, to give perspective.

255

The Horizontal and the Number Twelve

Twelve is easily divided by 2, 3,4, and 6, giving it mathematical order in keeping with that which is rational and logical. Traditionally, all numbers up to and including twelve represent the varied aspects of earthlife stages and cycles and the many levels of earthplane existence, whether visible or invisible, embodied or not.

The energy pattern of the twelve itself is said to establish and maintain the framework that provides a path toward completion of any given phase of life. It supposedly holds together in harmony and in balance that which populates and celebrates the circle of life. Time, space, and matter are understood to take on form and shape within the horizontal folds and storied layers of this magical energy, that of the twelve.

Key phrases: Numbers up to and including twelve— "The play of Life upon Itself." The twelve itself—"Graduation."

Key signs:

12 disciples of Jesus Christ

12 sons of Jacob

12 companions of Odysseus

12 princesses of Medea

12 shepherds of Romulus

12 peers of Roland

12 dismembered body parts of Osiris

12 knights of King Arthur's Round Table

12 gods of Mount Olympus

12 tribes of Israel

12-tribe system of culture bearers

12 equal divisions of the Wheel of Life

12 churches of revelation

12 signs of the zodiac

12 mysteries of life

12 colleges or schools the soul must pass through in its earthly sojourns

12 challenges of soul growth

12 powers of karma

12 hours of night after death

12 hours of day after birth

12 rays of the light spectrum

12 harmonics of sound

12 levels to heaven

12 levels in hell

12 months of the calendar

12 is the mystical number for the sun

"As above, so below" best exemplifies the outworking of twelve's horizontal energy—that organic balance between intellect and intuition, law and idealism, boundaries and independence, civilization and nature.

The Vertical and the Number Thirteen

Thirteen cannot be divided by a number below itself, making its connections beyond that which is rational or orderly. The energy pattern of the thirteen itself is said to guarantee a shift in octaves of vibration after the completion of any given life phase. It supposedly enables life-forms to transform and transmute in a continual process of evolution or devolution. Time, space, and matter are understood to lose form and shape within the vertical convergence and suspended animation of this mystical energy, that of the thirteen.

Key phrase: "The breath of Life breathing."

Key signs of *devolution:*

13th Samskara (important ritual) in Hinduism, one's death ceremony

13th card in Tarot deck, death

13th stage of life in ancient Egypt, death

$1 + 3 = 4$, and 4 is the number of death in Japanese

13th seat of honor at certain Babylonian religious festivals; person is executed as part of the ceremony

13th century, the Inquisition began in Europe; lasted three hundred years and caused the deaths of millions, the majority of which were women who practiced midwivery, home healing, and traditional goddess-oriented religions

13th guest, the uninvited bearer of strife and evil

13 steps to the gallows platform

13 coils of rope in the hangman's noose

13th day when on a Friday, a time of portent and doom

13th number, bad luck and misfortune

13th generation of Americans born since our republic's founding (between 1961 and 1981), the star-crossed or "lost" Generation X. (Imprinted with the need for transformation and transmutation, theirs is the challenge of bringing the energy shift of The Vertical to society as a whole—either for its detriment or renewal)

13th Apollo space flight, launched at the 13th minute of the 13th hour, then two days later, on April 13th, had an oxygen tank rupture, which put the mission at risk and threatened the lives of its crew (chronicled in a book and major motion picture, both entitled *Apollo 13*)

Keys signs of evolution (which also includes Generation X):

13 is the mystical number for Jesus Christ (12 + 1)

13 as a reference to the tabernacle itself, which held the Ark of the Covenant

13 items necessary for the tabernacle

13 as a sanctified number in ancient Israel

13th seat at King Arthur's Round Table, reserved for the knight who would one day find the Holy Grail

13, the number in pre-Columbian society for regeneration

13 cycles of the moon each year

13 menstrual cycles each year for each woman (said to be why 13 females make up a coven in Wicca—representative of blood, fertility, and lunar potency)

13 is the first of the teen years, puberty

13 years of age, both males and females are eligible to attain adulthood in most tribal communities

13 colonies instrumental in creating the American experiment in democracy

13 stripes on the U.S. flag

13 tail feathers on the national emblem of the United States, the eagle, as it appears on the silver dollar

13 as the predominant number throughout the Great Seal of the United States; stars, clouds around the stars, stripes, arrows, leaves, berries in olive branches, feathers in eagle's tail, layers of stones on pyramid, plus the number of letters in "*E Pluribus Unum*" and "*Annuit Coeptus*"—all total 13 each

13th amendment to our Constitution, abolishes slavery

13 Western states, each a bastion of independence and individualism

13 tectonic plates forming the earth's outer crust (was 12 until the recent split in the Indo-Australian plate just south of the equator)

13th Chapter of Psalms and Proverbs in the King James version of the Christian Bible, unusually meaningful; as well as the 13th chapter and the 13th verse of many New Testament books—especially First Corinthians: "And now abideth faith, hope, charity, these three; but the greatest of these is charity."

13 is the number in sacred geometry around which the famous labyrinth at Chartes Cathedral in France was designed (11 turns, center, plus outer edge for calendar keeping of lunar moons = 12 + 1, the Christ, or "everything is One")

13 is the mystical number for the moon

The "In-between" best exemplifies the outworking of thirteen energy—that "pivot point" The Vertical provides whereby we either ascend into the higher realms of spirit or descend into the lower realms of ego. Thirteen energy is the guarantor of change, not direction. Direction is determined by our readiness or lack of it to make the necessary shift in consciousness and embodiment required. Hence, stories about The Vertical are fraught with perils of unpredictability and what can be done to ensure safe passage.

As a brief aside, my own near-death encounter with The Vertical is covered in chapter 13, the chapter on the "In-between." An explanation of The Vertical appears here, in appendix IV. Both thirteen and four signify death in various cultures and traditions; The In-between" *is* The Vertical. I did *not* plan how the subjects in this book would be arranged; they simply flowed into place as I wrote.

Also, when I first composed chapter 13 on my computer, specifically when I was seeking to describe the cyclonic image I had met in death, what I had been typing abruptly bunched together on the terminal screen in such a manner that it formed an exact duplicate of the picture I had previously penned of the hourglass-shaped cyclones. The centerpoint and its rays enlarged in size as they seemed to head straight for me. Then the power of the centerpoint burst through the glass and exploded in my face. Radiation-like sparkles completely covered me and filled the room, along with a tinge-of-ammonia, flat-ozone smell. When I could move, I turned off the computer and unplugged it, picked up the terminal, put it in my car, and drove across town to the best computer specialists I could find. They were dumbfounded. Even though they claimed it was technically impossible for anything like this to happen, evidence to the contrary was incontrovertible. (They gave me a discount on a replacement; my only discomfort consisted of dancing sparkles in my vision for several days, rattled nerves, and a stinky office that required lots of fresh air.)

That incident underscored my convinction that I had indeed encountered The Vertical during my third near-death experience, and that I was on the right track in coming to terms with its purpose and meaning.

Appendix V

Just as understanding a person lets us realize why he acts the way he does, so each new degree of spiritual understanding lets us see why life acts the way it does.
—Craig Carter, D.D.

The case of Edward de Vere, seventeenth earl of Oxford, is a fascinating one. Not only does what is recorded of his life display obvious spurts in brain growth after what could have been three separate near-death experiences (maybe even four), but he may well be the ever-elusive but true author of the entire works of Shakespeare.

For the purpose of perspective, this appendix consists of an outline summarizing his life and times with incidents and a possible story line supplied by Leslie Anne Dixon, a professional research historian and expert on the subject. She recommends the following books as excellent sources for those who might want more detail about the enigma of Edward de Vere:

The Mysterious William Shakespeare. Ogburn, Charlton, Jr. (the son). New York: Dodd, Mead and Co., 1984.

Shake-Speare: The Real Man behind the Name. Ogburn, Charlton, Jr., and Dorothy Ogburn (mother-and-son team). New York: William Morrow and Co., 1962.

The Star of England. Ogburn, Charlton, and Dorothy Ogburn (husband-and-wife team). New York: Coward-McCann, 1952.

Dixon herself made international, front-page headlines when, in the spring of 1992, her research, along with that of her associate Lillian F. Schwartz, revealed that the most well-known image of "Shakespeare" is patterned *on the likeness of a woman*!

Published in *Pixel* magazine (vol. 3, no. 1, March/April 1992), Schwartz's article, "The Mask of Shakespeare," shows conclusively that the engraving done by Martin Droeshout, and published as the frontispiece of the First Folio of Shakespeare in 1623, is really a disguised copy of Queen Elizabeth I's 1588 official royal portrait by George Gower. By scanning, digitizing, and scaling both engraving and painting to size, then overlaying them and counting pixels (dots per inch), they discovered that face structure and contours between the two were not only similar, they were identical. Why the queen's portrait was "masculinized" or by whom is now being investigated.

Armed with documentation from this discovery plus over one hundred reference sources, Dixon submitted "Sweet Silent Signature," her article on the topic, for publication in *The American Scholar,* the official magazine of Phi Beta Kappa. Even though the editorial staff confirmed the accuracy of her details and references, her article was rejected. "It's too controversial," she was told.

What amazes me about Dixon's work and the possible story line that emerges from it is that it dovetails with my own study of historical figures for what might be examples of people who underwent a near-death experience and subsequent brain shift. Dixon located me, by the way, when, on another phase of her project, she was seeking someone who had researched the elder runes (hieroglyphs), which predate by thousands of years the development and use of Futhark (the Germanic/Viking system of hieroglyphic writing). I had combined my efforts in this area with "how-to-use-them" instructions in two books: *The Magical Language of Runes* (Bear and Co., 1990) and *Goddess Runes* (Avon Books, 1996).

Here is the outline as promised, summarized in before-and-after style so the possibility of a brain shift can be considered in conjunction with the actual death events Oxford survived.

Edward de Vere, Viscount Bulbeck (1550–1604), was born into the oldest noble family in England, tracing its roots to the time of William the

Conqueror. His father, John, the sixteenth earl, was late in years when his second wife, Margaret Golding, delivered a son (they had an older daughter). Unusually bright as a child, Edward was seventh in line to the throne during the reign of Edward VI through descent from the sister of Henry VIII.

- At age six, started composing poetry

- At age eight, outpaced all his tutors and began study at St. John's College

- At age nine, enrolled at Queen's College, Cambridge, and was translating Latin into English by age ten

- At age eleven, met Queen Elizabeth I and at her request composed the two-thousand-line poem "Romeus and Juliet" as a demonstration of his ability (later used as the prototype for the play *Romeo and Juliet*)

First Near-Death Event (*pay attention to what occurred afterward*): When Edward was twelve, his father suddenly collapsed and died. Young Edward fell ill at the same time (perhaps by drinking from the same cup as his father—though this cannot be proved), and hung at the edge of death for several weeks. Edward, now seventeenth earl of Oxford and lord great chamberlain of England, became a ward of the queen in London. Shortly after, he was plagued by reoccurring dreams or visions of his dead father, and was told by him that he had been murdered. No one believed the twelve-year-old's story of these ghostly visitations. Within two months, his mother remarried—to Charles Tyrall, the man who might have poisoned his father (Tyrall was an agent for the powerful earl of Leicester, a favorite of Elizabeth's). The marriage both shocked and grieved Edward. Although he inherited vast estates, several titles, and immense wealth, his elder half-sister tried to thwart his claim (and failed). Then his guardian, Baron Burghley, and Leicester (who were appointed to oversee the inheritance) greatly diminished it. Between coping with his father's death, his own near-death, the corruption and political maneuverings of the Queen's Court at too early an age, not to mention his mother's remarriage, he sank into a deep depression (as Sonnets 30 and 66 express).

- At age twelve, suddenly began to churn out compositions, publishing "Romeus and Juliet" under a pseudonym. Since noble men of his rank never published, he was severely chastised—a painful lesson he would not forget.

- At age fourteen, graduated from college with a bachelor's degree, his college deans noting the incredible jump in his intelligence.

- At age fourteen, his first play, a tragicomedy, was performed at court. When he was between the ages of fourteen and seventeen, the queen seduced him. ("Venus and Adonis" his version of the affair.) He fell in love with Elizabeth and with her wrote words and music to "When I Was Fair and Young."

- At age fifteen, learned that Leicester was also the queen's lover, a fact that overwhelmed him. He retreated into study and writing while haunted by dreams or visions of his dead father—they would go to "a cold gray place of mist" to talk. (Leicester was to become Claudius in *Hamlet,* Elizabeth would appear as Gertrude, and Burghley as Polonius.)

- At age sixteen, graduated with a master's degree, published a number of student plays under other names (much to the chagrin of his guardian, Burghley), and then enrolled in law school.

- At age sixteen, his capacity for learning accelerated even more; he read several dozen books per week in five languages (owned hundreds of books himself, including the Geneva Bible "Shakespeare" quoted from).

- At age seventeen, was still the queen's lover but now accepting of the secrecy demanded to maintain the "Virgin Queen" myth. Took up jousting and was unbeatable. Had a reputation for being totally fearless and without any regard for death or his own safety.

- At age seventeen, published continuously. Wrote dozens of pieces while at Grays Inn (including *Horestes,* a first draft of *Hamlet* and the Robin Hood Quadrology in honor of his ancestor—*King John* being part two of that set). Accidentally killed a man sent to spy on him by his guardian, stabbing him through a curtain. Was found innocent in court (the scene appears in his autobiography when Hamlet stabs Polonius through a curtain).

- At age nineteen, was recruited by Walsingham into Elizabeth's espionage network, and became a secret agent.

Second Near-Death Event (again note the abrupt change in Oxford afterward): A silent war with Leicester may have reached an apex. Oxford believed that his father and he were poisoned, but he could not go to the queen without evidence. At nineteen, he fell mysteriously ill after having dinner with Leicester, and once again hung at death's door for several weeks. He improved after the queen moved him to a safe place, where her own physician could attend him. It took a year for him to recover, whereupon he begged her for a commission in the army. He had finished possibly as many as eighteen plays before he left for the North Rebellion on the Scottish border to serve under his uncle and foster father, the earl of Sussex. Lacking any fear of death, he constantly risked his life to save others (was mentioned in dispatches for bravery several times). While stationed with the garrison at Glamis Castle, he heard about the legend of MacBeth, the Usurper. He literally overlaid his own story atop the local one to produce *MacBeth*, a drama that enabled him to express visions of the future that he could have seen during his second near-death experience (along with life-and-death issues recently faced). He refused to live with his guardian after returning from the war. He was accused of treason in aiding his cousin, the condemned Catholic duke of Norfolk, until the queen stepped in and rescued him, forcing him to agree to a loveless but safely Protestant marriage of convenience with Burghley's daughter, Anne (echoed in the relationship between Hamlet and Ophelia). While remaining the queen's lover, he switched from soldier to scholar to produce with John Dee the first book written in English on astrology. Then he wrote the first English book on navigation using astronomy.

- At age twenty, his mother died (she had shown little love for him after her remarriage); her husband died shortly thereafter.

- At age twenty-one, took his place in the House of Lords and, as lord great chamberlain, carried the sword of state before the queen.

- At age twenty-one, won the Royal Tournament as the Knight of the Tree of Gold, defeating Elizabeth's own bodyguard.

- At age twenty-two, failed to appear for his own wedding; supposedly joined up with the Catholics in Brussels (actually on a spying mission for the queen). Reported his activities in poetry and plays. The wedding took place nonetheless. Indulged in many trips, frantic living, drunkenness—yet all the while his intelligence skyrocketed.

- At age twenty-two, was appointed a judge of the appeals court (on the side still a secret agent). Took up horticulture and became an expert on herbs and poisons. Continued to write, had his own ship (the *Edward Bonadventure*), and could sail it. Was one of the judges at the trial of Mary Queen of Scots. Fought endlessly but unsuccessfully against her death penalty.

- At age twenty-four, became a father when the queen gave birth to his son (her fourth pregnancy) probably on Midsummer Night's Eve, June 20, 1574. The two had been secretly legally betrothed when this happened, making the boy, Henry, according to the fine points of English law, the legitimate heir to the throne. Oxford's belief that Elizabeth would marry him was crushed when she spirited the child away to safekeeping and would not let Oxford know where he was (wrote *A Midsummer Night's Dream* about their falling out over the changeling boy raised elsewhere under another name—references to the sun in *Hamlet* often are about his son). Lapsed into deep depression, outbursts of wildness (showed in Henry IV Graves Hill incident), left for Europe to get away and to carry out missions—one in Spain.

- At age twenty-six, while in Europe, his wife-in-name-only became pregnant and was delivered of a daughter. Her father, Burghley, insisted the child was Oxford's to secure his estate should he die before returning home. However, Oxford, knowing different, refused to claim the child.

Third Near-Death Event: At the age of twenty-six, Oxford was badly hurt in an accident while in Italy. His knee was crushed and he nearly died of fever. Immediately after, he developed a sudden interest in Near East, Greek, and Roman cultures, visiting all of them. The glare of southern sunlight bothered him greatly, and he wrote home complaining of it (began *Merchant of Venus* during this period). Invigorated, he challenged the greatest knight in Europe, Don John, to a combat "to the death" (the man never showed up), and was captured by pirates (like Hamlet) on the way home. He returned to England to find himself the butt of jokes over his wife having had a child while he was gone. He shut himself off from people and from light, wore only black, and was considered suicidal. He became certain that the daughter his "wife" had borne was a case of incest (see references in *Hamlet*) for the purpose of seizing his estates. When he came out of his depression, he went into a fury of writing (wrote *Titus Andronicus* with its atrocities after hearing that Don John was responsible for the massacre at Antwerp). He then abruptly built the first of three playhouses in England so his plays could be publicly performed (gaining a reputation as an extreme eccentric in manners and dress). History plays poured from him, so much so that he recruited other writers and used their names for authorship. Under his secretary's name, he also wrote what is considered to be the first English novel, *Euphues, An Anatomy of Wit*, published in 1579. After nine years of not knowing anything about his son, he learned that young Henry had been made a ward of the queen. Father and son were finally reunited. The boy became Henry Wriothesley, third earl of Southampton (to whom the Fair-Youth sonnets are dedicated—love sonnets from father to the royal son he could never openly claim).

At thirty-one, Edward nearly died a fourth time, this time from a terrible sword fight (see *Romeo and Juliet*) that left him lame. Convinced he would not live, he finished *Hamlet* so his son would know something of his father's life. He accepted his estranged wife's offer to come and nurse him. Surprisingly, they fell in love, and afterward Anne bore him three daughters and a son who died quite young. During this time, his uncle, Sussex, died the same way his father had. Sussex warned Oxford with his last breath about Leicester and his poisonings. Not fully recovered from his own wounds, Oxford still commanded his own ship in the Battle of the Spanish Armada. The ship was crippled, and, while putting in for repairs, he had drinks with Leicester. Probably cups were switched, because Leicester died suddenly of the poison intended for Oxford. Thanks to a freak storm and speedy frigates, England was saved and the Spanish defeated, freeing Elizabeth from the "Virgin Queen" myth she had used from her earliest years to thwart an invasion of England. Oxford was bankrupted from embezzlements to his estates, travel, building theaters, and losing three fortunes trying to prove there was a "North West Passage" in the new world (he never did—a joke in *Hamlet*). His wife, Anne, died while he was at sea, so his father-in-law, Burghley, took his daughters to raise. Hearing this, the queen awarded him a new estate for his great service to the realm, and picked one of her ladies-in-waiting, Elizabeth Trentham, to be his second wife. Oxford and his new wife loved each other deeply and she bore him a son, the only legitimate male heir he could ever openly claim. He settled into writing his greatest works while a happy recluse and family man. Then Henry, tired of waiting to be recognized, became involved in the Essex Rebellion, was convicted of treason, and imprisoned, where he would be safe until Elizabeth needed him. She was left helpless by a stroke, however, and when she died, James VI of Scotland ascended the throne in the unrecognized Henry's place. (A silent war would be played out between James and Henry throughout their lives.) Oxford lost all heart and failed to appear in the House of Lords for the first time in his life, saying that, with the queen gone, he could no longer muster the will for it. He finished up the last of his plays (*The Tempest*) and died too, probably of the plague.

The early drafts of Shakespearean plays were performed in the Queen's Court *before* Shakespeare of Stratford was born. To understand the jokes, politics, references, and fashions of these plays, it is necessary to date them about fifteen years earlier, to Oxford's time, rather than in "Shakespeare's." The enrichment of language, most of the thirty-two-hundred words supposedly "invented" by Shakespeare, appear in Oxford's correspondence *after his near-death episode at age twelve*! The works of Shakespeare take English from an agricultural language to a lively, colorful parade of new concepts and new sounds and abstracts of soul, pride, and patriotism. Shakespeare of Stratford was never paid a penny in royalties, yet he always received an allotment of one-thousand pounds (1 percent of the national budget) from an unknown benefactor. He was unschooled and went unprosecuted for his many indiscretions. There is little doubt that he was used as a means of concealment for authorship.

And there is no doubt that Elizabeth used Oxford as well, shamelessly, for the benefit of England. She knew that, through him, she could maneuver her subjects to think and act as she wished, fooling her enemies in the process. England was virtually bankrupt when she became queen. Her quandary: How does one inspire a weakened people? The "Virgin Queen" ruse kept other rulers guessing, which gave her the time she needed to build an economy, an army, and a navy. Through Oxford, she had a way to reach the public directly. In this manner, her subjects learned English history, developed pride in citizenship, and were whipped into war fever in time to meet the Armada threat. How she manipulated Oxford is plain enough. What isn't readily known is that Elizabeth herself nearly died as a youngster and displayed the pattern of a brain shift afterward, wearing out every tutor she was given and becoming proficient in five languages. She recognized the same qualities she possessed in Oxford. Indeed, they were often said to be twin minds. (Dixon, who pointed this out to me, has had multiple near-death incidents throughout her life and experienced a brain shift after each one.)

Although it can be argued that the life and times of Edward de Vere, seventeenth earl of Oxford, were shaped by many forces, I believe that the near-death phenomenon was prominent among them and especially the evidence suggestive of a brain shift. (Note the lack of fear plus extreme sensitivity to bright sunlight—both typical to the aftereffects of near-death and subsequent brain shift.) Should you wish to reach Leslie Anne Dixon concerning her research, write in care of the following address and your correspondence will be forwarded: You Can Change Your Life, P.O. Box 7691, Charlottesville, VA 22906-7691. Put "Attention—Dixon" on the lower left corner of the envelope.

A Heaven We Can Seek Together
by Stephen E. Hanson

Many have rode out gallantly;
Solitary, single, and alone;
Hoping to be the only one
To find some sword, or grail, or stone.

Many have, without companion,
Made pilgrimage both arduous and long;
Having as virtue for their vision
The boldness of the strong.

Many have gone into hermitages
Wearing sandals and shirts of hair,
Journeying on the road within
To a fleshless world more fair.

Many have scaled mountains by themselves,
Seeking victories purer than air,
At the end of resource, at the edge of life,
To be free from fear and care.

Let them go. Each has his path.
May to them the force be kind.
I bless them on their separate ways.
Let what they seek be what they find.

But different is our sacred journey,
One to a warm and human land.
There is that which can't be won
Unless we travel hand in hand.
A heaven we can seek together
Is not the same one sought apart.
Instructions written for love's quest
Are in the language of the heart.

Resources

Trust, but verify.
—Russian Proverb

Socrates once said, "the unexamined life is not worth living." While altering realities and states of consciousness, it becomes imperative to know thyself. Once your brain shifts, you expand. When that happens, responsibilities increase, not decrease, and you become an active participant in the very fabric of existence. The only asset you can bring to this transformation is what you have "builded" yourself to be. What you truly are, then, determines the consequences of the brain shift you make.

And we are challenged to make this shift. In the Christian Bible, Romans 12:2 says, "be not conformed to this world: but be ye transformed by the renewing of your mind, that ye may prove what is that good, and acceptable, and perfect, will of God."

To aid in renewing your mind, not only have I designed a book that can simulate the various stages of a brain shift just by being read, but I have also prepared this section of specific and pertinent information arranged by subject matter to offer next steps should you choose to accept the challenge of change. Resources listed here differ from those in *Beyond the Light* (Carol Pub Co., 1994), as the material in this book is broader ranging and more tantalizing and farther afield. Coming to terms with "self" actually necessitates that we rethink what constitutes the whole of life.

By order of relevance, topics covered are

This section is not meant to comprise a comprehensive listing, nor can I make any promises or guarantees about anything or anyone so presented. I simply offer this material to you as a gesture of sharing. What you do with it depends on you.

Be certain to visit my website at *www.pmhatwater.com*, and subscribe to my free monthly newsletter. To receive a brochure about my offerings, send a stamped, self-addressed envelope to You Can Change Your Life, P. O. Box 7691, Charlottesville, VA 22906-7691.

The Near-Death Phenomenon and The Other Side

The International Association for Near-Death Studies (IANDS) exists to impart knowledge concerning near-death experiences and their implications, to encourage and support research dealing with the experience and related phenomena, and to aid people in starting local groups that desire to explore the subject. They have numerous publications, among them the scholarly *Journal of Near-Death Studies*, a general-interest newsletter *Vital Signs*, and various brochures every hospital and clinic on earth should have on hand. Membership in this nonprofit organization is open to anyone; dues are annual. Reportings of near-death experiences are solicited and inquiries are welcome. Ask for their list of national and international chapters should you be interested in visiting any of them.

To contact IANDS or to make a contribution to the Near-Death Experience Research Fund, contact IANDS, 2741 Campus Walk Avenue, Bldg. 500, Durham, NC 27705; (919) 383-7940; *services@iands.org; www.iands.org.*

A few suggestions for information about death and the afterlife follow.

The Death Process

As You Die. An audio presentation by P.M.H. Atwater, L.H.D., that talks a dying individual through death as it physically occurs and the separation of the soul. Available from the author at *www.pmhatwater.com* or from Focus Worldwide Network, 229 N. Vermont Street, Covington, LA 70433; (985) 635-0333; *www.focustvonline.com.*

Final Gifts: Understanding the Special Awareness, Needs, and Communications of the Dying. Maggie Callanan and Patricia Kelley. New York: Simon & Schuster, 1992.

Light on Death: The Spiritual Art of Dying. Phillip Jones. San Rafael: Mandala Publishing, 2007.

Mandalas: Vision of Heaven and Earth and *The Human Journey.* Mirtala. Two videos, each focusing on scenes of transformational sculpture set to music, which are especially helpful in hospice and for the dying. Available from Mirtala, 10780 E. Oak Creek Valley Drive, Cornville, AZ 86325; *mirtala@earthlink.net.*

On Death and Dying: What the Dying Have to Teach Doctors, Nurses, Clergy, and Their Own Families. Elisabeth Kübler-Ross, M.D. New York: Macmillan, 1993.

Parting Visions: Uses and Meaning of Pre-Death, Psychic and Spiritual Experiences. Melvin Morse, M.D., with Paul Perry. New York: Villard Books, 1994.

The Other Side

Afterlife Encounters: Ordinary People, Extraordinary Experiences. Dianne Arcangel. Charlottesville, VA: Hampton Roads, 2005.

The Afterlife Experiments: Breakthrough Scientific Evidence of Life After Death. Gary E. Schwartz, Ph.D., with William Simon. New York: Atria Books, 2003.

After We Die, What Then? Answers to Questions about Life After Death. George W. Meek. Kingston, RI: Metascience Foundation, 1980. Available from Mark Macy at Continuing Life Research, P.O. Box 11036, Boulder, CO 80301; (303) 673-0660.

Evidence of the Afterlife: The Science of Near-Death Experiences. Jeffrey Long, M.D., with Paul Perry. San Francisco: HarperOne, 2010.

Glimpses of Eternity: Sharing a Loved One's Passage from This Life to the Next. Raymond Moody, M.D., Ph.D., with Paul Perry. Nashville: Ideals, 2010.

Heaven and Hell. Emanuel Swedenborg. West Chester, PA: Swedenborg Foundation, 1984.

Life Between Life. Joel L. Whitton, M.D., Ph.D., and Joe Fisher. New York: Warner Books, 1986.

Reunions: Visionary Encounters with Departed Loved Ones. Raymond Moody, M.D., Ph.D., with Paul Perry. New York: Ballantine Books, 1993.

We Live Forever: The Real Truth About Death. P.M.H. Atwater, L.H.D. Virginia Beach, VA: A.R.E. Press, 2004.

Health Issues

In a sermon given during a healing mission back in the summer of 1976 at St. Paul's Church by the Lake, Chicago, Illinois, Rev. Alfred Price made a stunning statement that presaged the medical discovery of mind-body links to illness. "Every word you speak requires the harmonious use of more than 72 muscles. Good health is not, and can never be, merely the absence of disease. Good health is the harmonious operation of every single organ of the body, every thought of the mind, and every emotion of the heart."

With this in mind, the U.S. government has finally recognized the need to investigate alternative and complementary techniques in medicine and has established the Office for the Study of Unconventional Medical Practices, at the National Institutes of Health. Four states have now legalized alternative medicine: North Carolina, Alaska, Washington, and New York. This issue is extremely important for those who experience a brain shift, as they often lose their ability to tolerate pharmaceuticals afterward.

Here are some groups and resources that are making a difference:

American Holistic Medical Association, 27629 Chagrin Blvd., Suite 213, Woodmere, OH 44122; *www.holisticmedicine.org.*

Center for Mind-Body Medicine, 5225 Connecticut Avenue NW, Suite 415, Washington, DC 20015; *www.cmbm.org.*

Center for Science in the Public Interest, 1220 L Street N.W., Suite 300, Washington, DC 20005; *www.cspinet.org.*

Institute of HeartMath, 14700 West Park Avenue, Boulder Creek, CA 95006; *www.heartmath.org.*

Natural Health Magazine, 17 Station Street, Box 1200, Brookline, MA 02147; *www.naturalhealthmag.com.*

Prevention Magazine, 400 South Tenth Street, Emmaus, PA 18098; *www.prevention.com.*

Special Video

Energy Anatomy and Self-Diagnosis: The Language of Your Body's Power Centers is a three-hour, two-tape video of a workshop with Caroline Myss, a medical intuitive and co-author with C. Norman Shealy, M.D., of Creation of Health. Produced by MHMH Productions, the tape set should be available from your favorite bookstore, or through Myss at 1210 Hirsch Street, Melrose Park, IL 60160.

Books on the Subject

Anatomy of an Illness. Norman Cousins. New York: Bantam Doubleday, 1983.

Cross Currents: The Perils of Electropollution, The Promise of Electromedicine. Robert O. Becker, M.D. New York: Jeremy P. Tarcher / Penguin, 1990. A classic.

Decoding the Secret Language of Your Body. Martin Rush, M.D. New York: Fireside, 1994.

Energy Medicine: Balancing Your Body's Energies for Optimal Health, Joy, and Vitality. Donna Eden with David Feinstein, Ph.D. New York: Jeremy P. Tarcher / Penguin, 2008.

Healing and the Mind, Bill Moyers. New York: Doubleday, 1993.

Healing Words: The Power of Prayer and the Practice of Medicine. Larry Dossey, M.D. San Francisco: HarperOne, 1995.

Homeopathic Self-Care: The Quick & Easy Guide for the Whole Family. Robert Ullman, N.D., and Judyth Reichenberg-Ullman, N.D. New York: Three Rivers Press, 1997.

Prescription for Natural Cures: A Self-Care Guide for Treating Health Problems with Natural Remedies Including Diet, Nutrition, Supplements, and Other Holistic Methods. James F. Balch, M.D., Mark Stengler, N.M.D., and Robin Young Balch, N.D. Hoboken, NJ: Wiley, 2011.

Prescription for Nutritional Healing: A Practical A-to-Z Reference to Drug-Free Remedies Using Vitamins, Minerals, Herbs & Food Supplements. Phyllis Balch, C.N.C. New York: Avery Trade, 2010.

Brain/Mind, Consciousness, Memories

There is such a thing now as brain-gym workout centers (brain saloons), and they are popping up from Amsterdam to Seattle. These are places where people who want to get a psychedelic experience without ingesting drugs use high-tech, virtual-reality equipment to send themselves "into the stars." Sometimes called "mechanical meditation," these centers underscore how people are willing to risk the reduction of attention span and sensory sensitivity to experience quick escapes.

As far as I am concerned, brain/mind research is still in its infancy. The deeper researchers probe, the more surprises they find.

Sources on Brain/Mind and Consciousness

The Brain That Changes Itself. Norman Doidge, M.D. New York: Viking Penguin, 2007.

Catching the Light: The Entwined History of Light and Mind. Arthur Zajonc. New York: Bantam Books, 1993.

Cosmic Consciousness. Richard Maurice Bucke, M.D. Philadelphia: Innes and Sons, 1901. Currently by Citadel Press, New York City (continuous printing). A classic.

Creating Minds. Howard Gardner. New York: BasicBooks, 1993.

Entangled Minds: Extrasensory Experiences in a Quantum Reality. Dean Radin, Ph.D. New York: Paraview Pocket Books, 2006.

The High Performance Mind. Anna Wise. New York: Jeremy P. Tarcher / Putnam, 1997.

The Holotropic Mind: The Three Levels of Human Consciousness and How They Shape Our Lives. Stanislav Grof, M.D., with Hal Zina Bennett, Ph.D. San Francisco: HarperOne, 1993. A classic.

On the Nature of the Psyche. Carl Jung. San Rafael, CA: R. F. Hull and G. Adler, 1996.

Roots of Consciousness. Jeffrey Mishlove. New York: Random House, 1975. A classic.

An Important Book on Synesthesia

The Man Who Tasted Shapes: A Bizarre Medical Mystery Offers Revolutionary Insights into Emotions, Reasoning, and Consciousness. Richard E. Cytowic, M.D. New York: Jeremy P. Tarcher / Putnam, 1993.

An Important Book on the Link between Emotion and Memory

Descartes' Error: Emotion, Reason, and the Human Brain. Antonio R. Damasio. New York: Putnam, 1994.

Books on Lost Memories/False Memories

The Myth of Repressed Memory: False Memories and the Accusations of Sexual Abuse. Elizabeth F. Loftus and Katherine Ketcham. New York: St. Martin's Press, 1994.

Making Monsters: False Memories, Psychotherapy, and Sexual Hysteria. Richard Ofshe and Ethan Watters. New York: Scribner, 1994.

Unchained Memories: True Stories of Traumatic Memories, Lost and Found. Lenore Terr, M.D. New York: BasicBooks, 1994.

Child Development/Learning

The premier researcher and commentator on child development and the brain is Joseph Chilton Pearce. His books are classics, his work legendary. To understand what is happening to children's minds today and why, study the work of Pearce.

Some of his best:

The Biology of Transcendence. Rochester, VT: Park Street Press, 2002
The Bond of Power. New York: E.P. Dutton, 1981.
Evolution's End. San Francisco: HarperOne, 1993
Exploring the Crack in the Cosmic Egg. New York: Pocket Books, 1982.
The Heart-Mind Matrix: How the Heart Can Teach the Mind News Ways to Think. Rochester, VT: Park Street Press, 2012.
Magical Child. New York: Bantam Books, 1981.
Magical Child Matures. New York: Bantam Books, 1986.

Other Works about Child Development and Learning

Babies Remember Birth. David Chamberlain, Ph.D. Los Angeles: Jeremy P. Tarcher, 1988. (Also look for *Babies Know More Than You Think* by David Chamberlain, Ph.D., and Suzanne Arms.)
Children of the Fifth World: A Guide to the Coming Changes in Human Consciousness. P.M.H. Atwater, L.H.D. Rochester, VT: Bear & Company, 2012.
For Your Own Good: Hidden Cruelty in Child-Rearing and the Roots of Violence. Alice Miller. New York: Farrar, Straus & Giroux, 1983.
Frames of Mind: The Theory of Multiple Intelligences. Howard Gardner. New York: BasicBooks, 1983. A classic.
Geography of Childhood: Why Children Need Wild Places. Gary Paul Nabhan and Stephen Trimble. Boston: Beacon Press, 1994.
Little Big Minds: Sharing Philosophy with Kids. Marietta McCarty. New York: Jeremy P. Tarcher / Penguin, 2006.
Upside-Down Brilliance: The Visual-Spatial Learner. Linda Kreger Silverman, Ph.D. Denver, CO: Deleon Publishing, 2002.

Children's Spiritual Development

Angels Over Their Shoulders: Children's Encounters with Heavenly Beings. Brad Steiger and Sherry Hansen Steiger. New York: Fawcett Columbine, 1995.

Gently Lead: Or How to Teach Your Children about God While Finding Out for Yourself. Polly Berrien Berends. New York: HarperCollins, 1991.

The Secret Spiritual World of Children: The Breakthrough Discovery That Profoundly Alters Our Conventional View of Children's Mystical Experiences. Tobin Hart, Ph.D. Novato, CA: New World Library, 2003.

Co-Creating with Teenagers

Creative Rebellion: Positive Options for Teens is the title of both an exciting book and a workshop experience in the nineties, from Daniel Shahid Johnson. Calling himself "a fellow traveler who shares," he combines various cathartic personal awareness techniques to help teens build integrity and a sense of self-worth. His book is available at *www.amazon.com*.

"Midway Center for Creative Imagination" was initiated by David Oldfield as a way to present "The Journey," a self-discovery program for teenagers. The program combines the appeal of fantasy role-playing games with shared group therapy to help young people find positive solutions to the "necessary crises of adolescence." To contact Oldfield, the center's director, or to seek further information, contact Midway Center for Creative Imagination, 2112 F Street NW, #404, Washington, DC, 20037, www.midwaycenter.com.

Self-Development/Self-Deception

Clearly, if you deny your dark side, you block your growth. A common defense mechanism most of us have is to pour our anger out onto those who remind us of ourselves or what we do not want to admit about ourselves. When we react to certain things out of all proportion, it is usually because they awaken in us whatever we have repressed or suppressed. Dr. Jana Klenburg, a psychotherapist, is quoted in a newspaper clipping as saying, "whenever I wake up in the morning raging with anger against somebody for no immediate reason, I instantly know that I am in fact angry with myself and projecting my wrath on him." Let's face it. Growth involves darkness. Since seeds cannot sprout without the dark, it behooves us to recognize that in order to move forward, there must first be that stillness of doubt, fear, confusion, surrender, suspension, expansion, creativity, and awakening, as with any colloidal state. Drugs such as Prozac have become a substitute for fixing what needs fixing in one's life. They sometimes delay growth, not aid it. To grow, you have to confront what you don't want to see—you have to risk. Carl Gustav Jung put it best: "One does not become enlightened by imagining figures of light, but by making the darkness conscious."

Some books along this line follow.

The Challenge of Self-Development

Core Transformation: Reaching the Wellspring Within. Connirae Andreas, Ph.D., and Tamara Andreas, Moab, UT: Real People Press, 1994.

The Fifth Agreement: A Practical Guide to Self Mastery. Don Miguel Ruiz and Don Jose Ruiz, with Janet Mils. San Rafael, CA: Amber-Allen Publishers, 2011.

The Four Agreements: Toltec Wisdom Collection. Don Miguel Ruiz. San Rafael, CA: Amber-Allen Publishers, 2008. (Three-book boxed set.)

Fully Present: The Science, Art, and Practice of Mindfulness. Susan L. Smalley, Ph.D., and Diana Winston. Philadelphia: Da Capo Press, 2010.

The Game of Life and How to Play It. Florence Scovel Shinn. New York: Simon & Schuster, 1986. A classic.

Homecoming: Reclaiming and Championing Your Inner Child. John Bradshaw. New York: Bantam Books, 1992.

Self Matters: Creating Your Life from the Inside Out. Phillip C. McGraw, Ph.D. New York: Free Press, 2003.

Self-Esteem Affirmations: Motivational Affirmations for Building Confidence and Recognizing Self-Worth. Louise L. Hay. Carlsbad, CA: Hay House, 1998. (Audio CD.)

The Challenge of Self-Deception

Avalanche: Heretical Reflections on the Dark Side and the Light. W. Brugh Joy, M.D. New York: Ballantine Books, 1990.

The Call of Spiritual Emergency. Emma Bragdon, Ph.D. New York: Harper & Row, 1990.

Face It and Fix It: A Three-Step Plan to Break Free from Denial and Discover the Life You Deserve. Ken Seeley with Myatt Murphy. San Francisco: HarperOne, 2009.

Make Anger Your Ally: Harnessing Our Most Baffling Emotion. Neil Clark Warren, Ph.D. Garden City, NY: Doubleday, 1983.

Vital Lies, Simple Truths: The Psychology of Self-Deception. Daniel Goleman, Ph.D. New York: Simon & Schuster, 1985.

When Society Becomes an Addict. Anne Wilson Shaef. San Francisco: HarperOne, 1987.

Storytelling as a Resource for Growth

Awakening the Hidden Storyteller: How to Build a Storytelling Tradition in Your Family. Robin Moore. Boston: Shambhala, 1991.

Personal Mythology: Using Dreams, and Imagination to Discover Your Inner Story. David Feinstein, Ph.D., and Stanley Krippner, Ph.D. Los Angeles: Jeremy P. Tarcher, 1990.

Sacred Stories: A Celebration of the Power of Stories to Transform and Heal. Charles Simpkinson and Anne Simpkinson, Eds. San Francisco: HarperOne, 1993.

Your Mythic Journey: Finding Meaning in Your Life through Writing and Storytelling. Sam Keen and Anne Valley-Fox. New York: Jeremy P. Tarcher/Putnam, 1989.

The Wisdom of Fairy Tales. Rudolf Meyer. Edinburgh: Floris Books, 1988.

An Organization Devoted to the Art of Storytelling

Says Jimmy Neil Smith, director of the International Storytelling Center: "We've moved from traditional performance, where only the professionals tell stories, to a more participatory event. It builds on what was already there, a recognition that we all have stories. We live our lives in a network of stories, and all stories matter." For more information about the art form of storytelling, contact the International Storytelling

Center, 116 West Main Street, Jonesborough, TN 37659; (423) 753-2171; *www.storytellingcenter.net.*

For Professionals

I know of no one who has developed a quicker or more accurate way to pinpoint a person's inner landscape than the late Joan Kellogg. She spent a lifetime exploring the interface between psychology and the arts, and in every possible counseling environment. A remarkable and innovative thinker, the system she developed for interpreting drawn Mandalas (called *Great Round of the Mandala*) enables clinicians and researchers to understand how archetypal images, symbols, and colors reflect the dynamics of self and psychological development. Among those carrying on her work are Michele Takei, Ph.D., and Frank Takei, Ph.D. Contact: MARI® Creative Resources, Inc., 2532 Albemarle Avenue, Raleigh, NC 27610; (919) 821-4222; *http://maricreativesources.com/aboutus.htm.*

Independent Thinkers, Independent Schools

The time has come to honor the creative mind. If we as a society continue to hamstring our innovative movers and shakers, not only will we suffocate individual initiative, but progress itself will also cease. I understand the need for replication of research findings, and there is a place for that—but not with the kind of bias that deems invention or dissent the work of the devil and therefore bogus or nefarious. In my own case, I had to leave the United States to obtain the kind of doctorate I needed in Spiritual and Psychic Studies. I did this at the International College of Spiritual and Psychic Studies, Spiritual Science Fellowship, Townhouse Centre, 1974 de Maisonneuve West, Montreal, Quebec, H3H 1K5, Canada; (514) 937-8359; *info@iiihs.org; www.iiihs.org.*

For a book about the creativity drain still applicable today, read *The Last Intellectuals: American Culture in the Age of Academe* (Basic Books, 2000) by Russell Jacoby.

Schools That Truly Encourage Creativity in Learning

Atlantic University, 215 67th Street, Virginia Beach, VA 23451; 1-800-428-1512; *registrar@atlanticuniv.edu; www.atlanticuniv.edu.*

iEARN, 475 Riverside Drive, Suite 540, New York, NY 10115; (212) 870-2683; Fax (212) 870-2672; *iearn@us.iearn.org*; *www.iearn.org*.

New Dimensions Radio and Tapes, P. O. Box 7847, Santa Rosa, CA 95407; *www.newdimensions.org*.

Prescott College, RDP Admissions, 220 Grove Avenue, Prescott, AZ 86301; *www.prescott.edu*.

Schumacher College, The Old Postern, Dartington, Totnes, Devon TQ9 6EA, United Kingdom; (011) 44 1803-865934; Fax (0) 1803-866899; *admin@schumachercollege.org.uk*; *www.schumachercollege.org.uk*.

Union Institute & University, 440 East McMillan Street, Cincinnati, OH 45206-1925; 1-800-486-3116 or (513) 861-6400; Fax (513) 861-3218; *admissions@tui.edu*; *www.tui.edu*. You can create your own curriculum.

A Thought

> *Mediocrity is self-inflicted; genius is self-bestowed.*
> —Walter Russell

Integrity and Life Patterns

In *The Day America Told the Truth*, by James Patterson and Peter Kim (Prentice Hall, 1991), it was revealed that 91 percent of the American public admit to lying on a regular basis. This alarming fact can be kept in perspective if one studies the way business and government function in today's marketplace. What follows are some insightful offerings on the subject of lies, betrayal, virtue, and trust.

Adam Smith's Mistake: How a Moral Philosopher Invented Economics and Ended Morality. Kenneth Lux. Boston: Shambhala, 1990.

Dictatorship of Virtue: Multiculturalism and the Battle for America's Future. Richard Bernstein. New York: Knopf, 1994.

Forgiving the Unforgivable: The True Story of How Survivors of the Mumbai Terrorist Attack Answered Hatred with Compassion. Master Charles Cannon with Will Wilkinson. New York: Select Books, 2012.

The Moral Sense. James Q. Wilson. New York: Free Press, 1993.

Narcissism Epidemic: Living in the Age of Entitlement. Jean M. Twenge, Ph.D and W. Keith Campbell, Ph.D. New York: Free Press, 2010.

The Shadow Effect: Illuminating the Hidden Power of Your True Self. Deepak Chopra, Debbie Ford, and Marianne Williamson. New York: Harper-Collins, 2011.

Changing How We Do Business

Dare to Care. Louis Bohtlingk. New York: Cosimo, 2011. A book that describes a revolutionary system of generating and handling money and finance.

The Future of Money. Bernard Lietaer. New York: Random House, 2001.

The Mesh: Why the Future of Business Is Sharing. Lisa Gansky. New York: Portfolio Penguin, 2010.

Natural Capitalism: Creating the Next Industrial Revolution. Paul Hawken, Amory Lovins, and L. Hunter Lovins. London/Washington, DC: Earthscan, 2010.

Priceless: The Myth of Fair Value (and How to Take Advantage of It). William Poundstone. New York: Hill and Wang, 2011.

The Third Industrial Revolution: How Lateral Power Is Transforming Energy, the Economy, and the World. Jeremy Rifkin. New York: Palgrave Macmillian, 2011.

Keeping the Human Condition in Perspective

The Power of Myth. Joseph Campbell with Bill Moyers. New York: Anchor Books, 1991. A classic.

Spirituality and the Soul

Throughout the whole of history, there have always been avatars (great teachers or messiahs) who have sought to enlighten a deluded humankind about the divine potential present within each of us, and of our connection by right of our soul with the Godhead. Invariably, these people have been deemed insane or ignored or killed. The Western tradition of *Christos* (a term of Egyptian roots popularized in pre-Christian Greek mysticism but not associated with Jesus during his lifetime) defines the

Christ Consciousness as the result of a quickening and energizing force, which, once it has entered an individual's body, brings to full memory the knowledge of creation and the truth of God. This quickening (brain shift) was highly sought after, and schools sprang into existence that would teach people how to invoke the experience. One of these mystery schools initiated a sect called the Essenes with the foreknowledge that eventually from its ranks a new avatar would be born, one who would fulfill the ancient prophesies and herald a new age of enlightenment. Thus, Christ Jesus became important, not so much because he was human like the rest of us, but because he reminded us that *we are gods like him*. His how-to-do-it model remains as valid and effective today as it was several thousand years ago.

The idea of *worthship* then, the worthiness of the individual, is the real heart of both religion and spirituality. And each individual is worthy by fact of birth, part of the one true essence, the Greater Indivisible Whole. God is love, and God's love is inclusive, *not* exclusive, and people need to know this. There are many places where one may learn of religious viewpoints, but spiritual and metaphysical concepts have no such organizational structures. This confuses beginners. For that reason, I am including a rather lengthy listing of some outstanding publications in the field.

Cautionary and Informative Religious and Spiritual Overviews

The Case for God. Karen Armstrong. New York: Knopf, 2009.

Gospel Light: An Indispensable Guide to the Teachings of Jesus and the Customs of His Time. George M. Lamsa. San Francisco: HarperOne, 1986.

The Guru Papers: Masks of Authoritarian Power. Joel Kramer and Diana Alstad. Berkeley, CA: Frog Books, 1993.

Our Religions: The Seven World Religions Introduced by Preeminent Scholars from Each Tradition. Arvind Sharma, Masao Abe, Tu Wei-ming, Liu Xiaogan, Jacob Neusner, Harvey Cox, and Seyyed Hossein Nasr. San Francisco: HarperOne, 1993.

The Sins of Scripture: Exposing the Bible's Texts of Hate to Reveal the God of Love. John Shelby Spong. San Francisco: HarperOne, 2006.

Take Me for a Ride: Coming of Age in a Destructive Cult. Mark E. Laxer. College Park, MD: Outer Rim Press, 1993.

When God Becomes a Drug: Breaking the Chains of Religious Addiction and Abuse. Father Leo Booth. Los Angeles: Jeremy P. Tarcher, 1991.

Why God Won't Go Away: Brain Science and the Biology of Belief. Andrew Newberg, M.D., Eugene D'Aquili, M.D., Ph.D., and Vince Rause. New York: Ballantine Books, 2002.

Some Books on Spirituality

A Book of Saints: True Stories of How They Touch Our Lives. Anne Gordon. New York: Bantam Books, 1994.

The Christos. Vitvan. Baker, NV: School of the Natural Order, 1951. Available from School of the Natural Order, P. O. Box 150, Baker, NV 89311.

The Coming of the Cosmic Christ. Matthew Fox. San Francisco: HarperOne, 1988.

God Makes the Rivers to Flow: Sacred Literature of the World. Eknath Easwaran. Tomales, CA: Nilgiri Press, 1991.

God's Scripture: A Faithful Comparison, What Jews, Christians, and Muslims Must Know. Nader Pourhassan, Ph.D. Bloomington, IL: iUniverse, 2009.

Gratefulness, the Heart of Prayer: An Approach to Life in Fullness. Brother David Steindl-Rast. Mahwah, NJ: Paulist Press, 1984.

Hymns to an Unknown God: Awakening the Spirit in Everyday Life. Sam Keen. New York: Bantam Books, 1994.

Letters of the Scattered Brotherhood. Mary Strong, Ed. San Francisco: HarperOne, 1991.

The Prophet. Kahlil Gibran. New York: Alfred A. Knopf, 1923. Continuous printings. A timeless classic.

Shadows of the Sacred: Seeing through Spiritual Illusions. Frances Vaughan. Wheaton, IL: Quest Books, 1995.

Soul Writing: Conversing with Your Higher Self. Joanne DiMaggio. Charlottesville, VA: Olde Souls Press, 2011.

The Way of the Shaman. Michael Harner. San Francisco: HarperOne, 1980.

Wherever You Go, There You Are: Mindfulness Meditation in Everyday Life. Jon Kabat-Zinn. New York: Hyperion, 1994.

Wisdom Walk: Nine Practices for Creating Peace and Balance, from the World's Spiritual Traditions. Sage Bennet, Ph.D. Novato, CA: New World Library, 2007.

Of Special Interest

A Course in Miracles, as channeled through Helen Schucman with the help of Bill Thetford. A collection of challenging spiritual precepts and guidance. In bookstores or from Foundation for Inner Peace, P. O. Box 598, Mill Valley, CA; 94942-0598; *www.acim.org.*

The Love Principles are universal principles that appeared in luminescent form to Arleen Lorrance while deep in prayer, as she asked for guidance on how to handle unruly teenagers in her classroom. Lorrance, together with Diane Kennedy Pike, founded Teleos Institute, an organization offering a whole range of programs designed to deepen the spiritual experience and bring integrity back to the workplace. Inquire about the Love Principles or any of the programs offered: Teleos Institute, 7119 E. Shea Blvd., Suite 109, PMB 418, Scottsdale, AZ 85254-6107; *www. consciousnesswork.com.*

Nicholas Roerich Museum contains the largest collection outside of Russia of the mystical and religious paintings of Nicholas Roerich. Known internationally as an authority on religious icons, Roerich devised a way of handling color and brush strokes to convey stunning spiritual messages. Tours, books, reproductions. Contact: Nicholas Roerich Museum, 319 West 107th Street, New York, NY 10025-2799; *www.roerich.org.*

Notebooks of Paul Brunton is the culmination of thirty years of quiet from a successful businessman-turned-philosopher. Brunton developed a way to express spiritual insight through free-standing, numbered paragraphs. Available in single volumes or as a set. Contact the Paul Brunton Philosophic Foundation, 4936 NYS Route 414, Burdett, NY 14818; *www.paulbrunton.org.*

Science of Mind was written by Ernest Holmes in the 1920s and has become the core teaching for the Church of Religious Science (also known as United Centers of Spiritual Living). The text synergizes the best of religion and philosophy in one cohesive and readable volume. Also available in electronic format. Ask for information about the magazine *Science of Mind* if you are interested. Contact Centers for Spiritual Living, 573 Park Point Drive, Golden, CO 80401-7042; *www.scienceofmind.com.*

The Walking People is the first attempt by anyone to set the oral history of an entire lineage of indigenous people to words. Paula Underwood, an oral historian for *The Walking People*, used contemporary language to pass along her people's wisdom so "that the children's children's children may yet learn." The work is a rare treasure. Contact A Tribe of Two Press, P. O. Box 133, Bayfield, CO 81122; *www.tribeoftwopress.com.*

Expanded Awareness—Perception/Intuition

Various researchers in the field of consciousness studies claim there are forty-nine different ways of perceiving information and stimuli. Of that range, about 40 percent of our population is predominately visual in its method of perception, 40 percent predominately kinesthetic (feelers), and the remaining 20 percent auditory—*yet our entire education system is based on auditory learning, which bypasses 80 percent of our population.* No wonder, then, that so many individuals either do not know how to learn, are not interested in school, or find learning a bore. Remedies to this appalling situation are being tried, yet any remedy must address basic perceptual differences between people or it will not succeed.

And part of that difference in perceptual preference is *our intuitive faculties.* Intuition is a natural component of brain/mind functioning. There is nothing extra, super, or extraordinary about it. What society terms as psychic is merely an extension of faculties normal to us. Psychic abilities are considered suspect only by those who fail to recognize that anyone who is truly healthy *is automatically psychic.*

Some Books on Perception

Irreducible Mind: Toward a Psychology for the 21ˢᵗ Century. Edward F. Kelly, Emily Williams Kelly, Adam Crabtree, Alan Gauld, Michael Grosso, and Bruce Greyson. Blue Ridge Summit, PA: Rowman & Littlefield, 2007.

Margins of Reality: The Role of Consciousness in the Physical World. Robert G. Jahn and Brenda J. Dunne. Orlando, FL: Harcourt Brace & Company, 1987.

The Reality Illusion. Ralph Strauch, Ph.D. Barrytown, NY: Station Hill Press, 1989.

Some Books/Publications on Intuition and Psychic Abilities

Atlantis Rising Magazine, P.O. Box 441, Livingston, MT 59047; *www.atlantisrising.com.*

Extraordinary Knowing: Science, Skepticism, and the Inexplicable Powers of the Human Mind. Elizabeth Lloyd Mayer, Ph.D. New York: Bantam Books, 2008.

Fate Magazine, 170 Future Way, P. O. Box 460, Lakeville, MN 55044-0460; *www.fatemag.com.*

Joyful Evolution: A Guide for Loving Co-Creation with Your Conscious, Subconscious, and Superconscious Selves. Gordon Davidson. San Rafael, CA: Firebird Press, 2011.

Mind Trek: Exploring Consciousness, Time, and Space through Remote Viewing. Joseph McMoneagle. Charlottesville, VA: Hampton Roads, 1993.

Natural ESP: A Layman's Guide to Unlocking the Extra Sensory Power of Your Mind. Ingo Swann. New York: Bantam Books, 1987. A classic.

Sixth Sense: Unlocking Your Ultimate Mind Power, Laurie Nadel, Ph.D., with Judy Haims, and Robert Stempson. Lincoln, NE: ASJA Press, 2006.

Subtle Worlds: An Explorer's Field Notes. David Spangler. Everett, WA: Lorian Press, 2010.

Venture Inward, Hugh Lynn Cayce. New York: HarperCollins1985. A classic.

Venture Inward Magazine, A.R.E., 215 67th Street, Virginia Beach, VA 23451-2061; *www.edgarcayce.org.*

Individual and Group Offerings for Expanded Awareness

Dowsing. American Society of Dowsers, Danville, VT 05828-0024; (802) 684-3417; *www.dowsers.org.* Anyone from a four-year-old to grandparents can learn dowsing, and it's fun. Contrary to some notions, dowsing is readily teachable as a practical way to access octaves of vibration beyond what is normal for us. Emphasis on demonstration and results. Chapters throughout the country. Large conferences.

Intuitive Mind/Inner Knowing. Institute of Noetic Sciences, 625 2nd Street, Suite 200, Petaluma, CA 94952-5120; (707) 775-3500; *www. noetic.org.* IONS was founded by astronaut Edgar Mitchell and is a non-profit membership organization that supports the realization of our human potential. They sponsor leading-edge consciousness research, educational outreach, and engage a global learning community. International programs, retreats, activities.

Out-of-Body Experiences/Faculty Enhancements. Monroe Institute, P.O. Box 505, Lovingston, VA 22949; 1-800-541-2488; *www. monroeinstitute.org.* For over three decades this nonprofit institute has been offering gateway programs to explore altered states of consciousness. As a research facility, they are recognized internationally not only for the work of their founder, Robert Monroe, but also for documented technological breakthroughs in the field of brain and faculty enhancements. They are noted for classes in training people how to have out-of-body experiences. Full slate of ongoing activities.

Perceptual Models of Human Behavior. Neuro-Linguistic Programming (NLP), 24 E. 12th Street, New York, NY 10003; (212) 647-0860; *www.nlptraining.com.* The aim of NLP is to help people make successful communication a conscious choice, not a random event. Using maps of perceptual styles, the student learns to identify patterns/sequences of nonverbal behavior. NLP training centers are worldwide—they are recognized as leaders in perception and communication technology.

Psychic Development. Association for Research and Enlightenment (A.R.E.), 215 67th Street, Virginia Beach, VA 23451-2061; 1-800-333-4499; *www.edgarcayce.org.* Psychic development classes at the A.R.E. were designed by Henry Reed, Ph.D., and Carol Ann Liaros (former psychic trainer with the blind). The program is dynamic and focuses on internships, personal skills enhancement, annual conferences for professional psychics, plus a full range of research and experiential opportunities. Reed discovered that the psychology of intimacy (more than the biophysics of brainwave transmissions) governs telepathy and that true psychism is based on the heart-to-heart rapport that can exist between people.

Whole Brain/Whole Body Functioning. Educational Kinesiology Foundation, 1575 Spinnaker Drive, Suite 204B, Ventura, CA 93001; 1-800-356-2109; *www.braingym.org.* Developed by Paul E. Dennison, Ph.D., Edu-Kinesthetics utilizes muscle testing as a diagnostic and therapeutic tool to get in touch with natural body energy. The process offers a powerful way to deal with learning disabilities, dyslexia, and especially for broadening all manner of perceptual skills. Simple activities for whole brain learning, called Brain Gym, are available internationally for schools, individuals, and groups. Inquire about certified trainers.

References/Websites on Paranormal Research

A New Science of the Paranormal. Lawrence LeShan, Ph.D. Wheaton, IL: Quest Books, 2009.

The End of Materialism: How Evidence of the Paranormal Is Bringing Science and Spirit Together. Charles T. Tart, Ph.D. Oakland, CA: New Harbinger Publications, 2009.

The ESP Enigma: The Scientific Case for Psychic Phenomena. Diane Hennacy Powell, M.D. New York: Walker & Co., 2008.

Extraordinary Knowing: Science, Skepticism and the Inexplicable Powers of the Human Mind. Elizabeth Lloyd Mayer, Ph.D. New York: Bantam Books, 2008.

The Fundamentalist Mind: How Polarized Thinking Imperils Us All. Stephen Larsen. Wheaton, IL: Quest Books, 2007.

The Parapsychology Revolution: A Concise Anthology of Paranormal and Psychical Research. Robert M. Schoch, Ph.D., and Logan Yonavijak. New York: Tarcher / Penguin, 2009.

Science and Psychic Phenomena: The Fall of the House of Skeptics. Chris Carter. Rochester, VT: Inner Traditions, 2012.

Scientific Literacy and the Myth of the Scientific Method. Henry H. Bauer. Champaign, IL: University of Illinois Press, 1994.

About pseudo-skeptics: *www.debunkingskeptics.com.*

Scientists' Transcendent Experiences:
  http://psychology.ucdavis.edu/tart/taste/.

The Work of Rhea White and Exceptional Human Experiences:
  www.psi-mart.com (look for EHE publications).

Interspecies Communications

This deserves its own subject heading because of the importance of the topic. *You can communicate directly with nature and with other species of life forms on this planet.* It takes awhile to form the language skills necessary to accomplish this and to accept that it can be done, but once learned, the communication that results is nothing short of phenomenal.

Books on Nature Intelligence

The Attentive Heart: Conversations with Trees. Stephanie Kaza. New York: Fawcett Columbine, 1993.

Behaving as if the God in All Life Mattered. Machaelle Small Wright. Warrenton, VA: Perelandra, Ltd., 1983. From Perelandra Ltd., P. O. Box 3603, Warrenton, VA 20188.

Ecomysticism: The Profound Experience of Nature as a Spiritual Guide. Carl von Essen, M.D. Rochester, VT: Bear & Co., 2010.

Enchantment of the Faerie Realm. Ted Andrews. St. Paul, MN: Llewellyn, 1993.

The Findhorn Garden: Pioneering a New Vision of Humanity and Nature in Cooperation. The Findhorn Community. Forres, Scotland: The Findhorn Press, 1988. In U.S. A classic.

Gaia's Hidden Life: The Unseen Intelligence of Nature. Shirley Nicholson and Brenda Rosen, Compilers. Wheaton, IL: Quest Books, 1992.

To Hear the Angels Sing: An Odyssey of Co-Creation with the Devic Kingdom. Dorothy Maclean. Middletown, WI: Lorian Press, 1983. A classic.

Journey into Nature: A Spiritual Adventure. Michael J. Roads. Tiburon, CA: H. J. Kramer, 1987.

Partnering with Nature: The Wild Path to Reconnecting with the Earth. Catriona MacGregor. New York/Hillsboro, OR: Atria Books / Beyond Words, 2010.

Primary Perception: Biocommunication with Plants, Living Foods, and Human Cells. Cleve Backster. Anza, CA: White Rose Millennium Press, 2003. A classic.

The Secret Life of Plants. Peter Tompkins and Christopher Bird. New York: Harper Perennial, 1989. A classic.

Books on Animal Intelligence

Animal Communication: Our Sacred Connection. Jacquelin Smith. Lakeville, MN: Galde Press, 2005

Bats Sing, Mice Giggle: The Surprising Science of Animals' Inner Lives. Karen Shanor, Ph.D., and Jagmeet Kanwal, Ph.D. Minneapolis, MN: Consortium Book Sales/Icon Books, 2009.

Kinship with All Life. J. Allen Boone. San Francisco: HarperOne, 1976. A timeless classic.

Strange Powers of Pets. Brad Steiger and Sherry Hansen Steiger. New York: Berkley, 1993.

The Elephant Whisperer: My Life with the Herd in the African Wild. Lawrence Anthony with Graham Spence. New York: St. Martin's Griffin, 2012.

The Souls of Animals. Gary Kowalski. Walpole, NH: Stillpoint, 1991.

When Animals Speak: Advanced Interspecies Communication. Penelope Smith. Hillsboro, OR: Beyond Words Publishing, 1999.

Subtle Energies, Leading-Edge Sciences

In June 1990 Canadian Nobel Laureate John Polanyi delivered the commencement address at McGill University in Montreal. The following is quoted from that address. "At the heart of science lies discovery which involves a change in worldview. Discovery, in science or the arts, is possible only in societies which accord their citizens the freedom to pursue the truth where it may lead and which therefore have respect for different paths to that truth."

The academic censure that happened to Harold Saxton Burr, Ph.D., is still happening to others like him, and to alarming degrees. With all due respect for demonstration and replication, the simple fact is, *all mainstream science begins as heresy!* The saga of Gaston Naessens is a case in point. He invented a microscope that enables researchers to effectively challenge the work of Louis Pasteur and present findings in cellular microbiology. His microscope is so revolutionary that many feel it is worthy of a Nobel Prize—yet all he received was a jail sentence. (Reference: *The Life and Trials of Gaston Naessens: The Galileo of the Microscope* by Christopher Bird, 1990, from Les Presses, L'Universite de la Personne, St. Lambert, Montreal, Canada—in both English and French).

And then there's Hans Krebs who, in 1937, described a crucial energy-making reaction in living cells. His work was rejected and never received journal review, but he later won the Nobel Prize.

And Barbara McClintock, who, in 1950, discovered that genes jump around on chromosomes. Her peers refused to believe anything she said . . . until 1983 when she received the Nobel Prize for her work.

According to Harvard University science historian Everett Mendelsohn, "things which don't fit the reigning paradigm are pushed aside." Sorry if I seem to be adamant about this, but the fact is that over 70 percent of all medical procedures used today are unproven. Mathematics itself, even control-group testing, is not always right and cannot be completely verified scientifically. I, for one, have seen the truth of this with near-death studies, where personal bias on the part of researchers sometimes has more to do with acceptability than field observation and analysis.

What is often labeled as junk science can and many times does become the precursor to meaningful discovery. Considering the gravity of problems faced in our world today, the arrogance of a closed mind comes at a price that is much too high.

Subtle Energies/The Auric Field

Microscopic magnets have been found in the human brain. "They are little biological bar magnets made of crystals of the iron mineral magnetite," said geobiologist Joseph L. Kirschvink, of the California Institute of Technology. Homing pigeons, whales, salmon, honeybees, and some shellfish and bacteria possess microscopic magnets, too. This aids naviga-tion (and might explain why some people never get lost), but it also lends credence to the possible link between cancer and electromagnetic fields.

Even our very bodies act as passive antennas, boosting electrical sig-nals in the same way that holding onto television rabbit ears can clear a blurry picture. People who have undergone a brain shift often switch from passive to active, however, and become antennas that more directly affect and are affected by electromagnetics. Changes from the shift in brain chemistry and brain structure may account for this. Perhaps in some way, the natural magnetite is affected. Regardless of how or why, subtle energy realms suddenly become very real and very important to experiencers. And this includes the auric energy field within and sur-rounding all things (as well as the magnetic heart beat of the earth—the Schumann Resonance—which is the earth's natural magnetic field or bloodstream).

A leader in subtle-energy research and in identifying working with the human aura (electromagnetic field) is Barbara Ann Brennan, a former NASA physicist. Her two books are *Hands of Light: A Guide to Heal-ing through the Human Energy Field* (Bantam Books, 1988); and *Light Emerging: The Journey of Personal Healing* (Bantam Books, 1993). Bren-nan herself sees seven layers of energy around the body. She has given me permission to share her vision of these layers with you:

1. The first layer, close to the body, reveals physical sensations, our con-nection to earth, and our will to live. Looks like a web of tiny blue lines.

2. The second layer, the next layer out from the body, reveals our emo-tions and sensuality. Looks like a rainbow of colors (those most com-monly associated with the *chakras* or major endocrine centers).

3. The third layer out reveals our thoughts and mental processes. Appears mostly yellow with thought-form blobs of varying bright-ness and form.

4. The fourth layer out is related to our love of others and the world. Similar in color and form to the second, except has a rose tint.

5. The fifth layer is related to giving and receiving, "speaking your truth," and "connection to divine will." Looks like a photographic negative; is the blueprint of the physical body.

6. The sixth layer is related to spiritual ecstasy. Looks like shimmering light and is composed of pastel colors.

7. The last layer is an integration of spirituality and personality, the connection to God or to a creative life force. Its outer circumference is an egg shape around the body that contains all the other layers (as well as a grid structure of the body), and is composed of tiny threads of golden-silver light.

Should you wish to contact her or inquire about classes and workshops, contact Barbara Brennan School of Healing, 500 NE Spanish River Blvd., #208, Boca Raton, FL 33431; 1-800-924-2564; *www.barbarabrennan.com*.

Some Other Books on the Subject

The Chakras and the Human Energy Fields. Shafica Karagulla, M.D., and Dora van Gelder Kunz. Wheaton, IL: Quest Books, 1989.

Therapeutic Touch: How to Use Your Hands to Help or to Heal. Dolores Krieger, Ph.D, R.N. Englewood Cliffs, NJ: Prentice-Hall, 1979. A classic.

Wheels of Light: Chakras, Auras, and the Healing Energy of the Body. Rosalyn L. Bruyere. New York: Fireside, 1994.

Organizations on the Frontiers of Thought

Institute for Frontier Science, PMB 605, 6114 LaSalle Avenue, Oakland, CA 94611; *brubik@compuserve.com; www.healthy.net/frontierscience*. Beverly Rubik is the director.

International Society for the Study of Subtle Energies and Energy Medicine (ISSSEEM); all contacts now completely online at *www.issseem. org*. Check out their Links Section for a long list of organizations and researchers in this genre.

University of Science and Philosophy, P. O. Box 520, Waynesboro, VA 22980; Fax 330-650-0315; *www.philosophy.org*. Contains the work of the late Walter and Lao Russell. **Special Note:** *Scientific experimentation with* *state-of-the-art technology is used to test out Russell's original concepts. In a breakthrough in 1994, researchers were able to transmute nitrogen gas into helium-4 and lithium-5. Since lithium-5 has never been seen before, this discovery is of major importance. Further trials are pending.*

U.S. Psychotronics Association, 525 Juanita Vista, Crystal Lake, IL 60014; (530) 918-8772; *uspsychotronics@gmail.com*; *www.psychotronics.org.*

Books on Leading-Edge Science, Subtle Energies, and Consciousness

Chaos: Making a New Science. James Gleick. New York: Viking Penguin, 1987. A classic.

The Dark Side of the Brain: Major Discoveries in the Use of Kirlian Photography and Electrocrystal Therapy. Harry Oldfield and Roger Coghill. Salisbury, England: Element Books, 1991. In U.S.

The Field: The Quest for the Secret Force of the Universe. Lynne McTaggart. New York: Harper, 2008.

The Living Energy Universe: A Fundamental Discovery That Transforms Science and Medicine. Gary E. Schwartz, Ph.D., and Linda G.S. Russek, Ph.D. Charlottesville, VA: Hampton Roads, 2006.

Mind, Body & Spirit: Challenges of Science & Faith. William Pillow. Bloomington, IN: iUniverse, 2010.

One White Crow. George McMullen. Norfolk, VA: Hampton Roads, 1995. About psychic archaeology.

Science and the Akashic Field: An Integral Theory of Everything. Ervin Laszlo. Rochester, VT: Inner Traditions, 2004.

Stalking the Wild Pendulum: On the Mechanics of Consciousness. Itzhak Bentov. Rochester, VT: Destiny Books/Inner Traditions, 1988. A classic.

The Subtle Body: An Encyclopedia of Your Energetic Anatomy. Cyndi Dale. Louisville, CO: Sounds True, 2009.

Sacred Art and Architecture, Sustainability, Earth Energies

Sacred Art

Book on Fractal Patterns: *Fractals, the Patterns of Chaos: Discovering a New Aethetic of Art, Science, and Nature.* John Briggs. New York: Simon & Schuster, 1992.

Alfred Dolezal. Shangri La Studios, 4394 Garth Road, Charlottesville, VA 22901; (434) 823-6410; *www.alfreddolezal.com*. Creating art for inspiration and the inner journey, he is noted for precise detail in symbolizing aspects of higher concepts and universal law. A description of the painting's meaning accompanies each piece. Some reproductions appear in The Artist's Portfolio (a pictorial set of seventeen paintings) and in special calendars.

Alex Grey. Chapel of Sacred Mirrors, 46 Deer Hill Road, Wappingers Falls, NY 12590; (845) 297-2323; *www.cosm.org*. His mystical paintings articulate the realms of consciousness. Three of his books are:

Net of Being. Alex Grey with Allyson Grey. Rochester, VT: Inner Traditions, 2012. *Sacred Mirrors: The Visionary Art of Alex Grey*. Alex Grey with Ken Wilbur and Carlo McCormick. Rochester, VT. Inner Traditions, 1990. *Transfigurations*. Alex Grey. Rochester, VT: Inner Traditions, 2004.

Poetry

I consider poetry a sacred art form. With that in mind, I'd like to recommend the following people, two of whom are gifted poets, the other a leading commentator on the subject.

Bill Moyers: His book *The Language of Life* (Doubleday, 1995) is a celebration of poetry and was both a popular book and a television special on the Public Broadcasting System.

Gary Snyder: A Pulitzer Prize-winning poet and translator of Zen Buddhist texts. I have referred to him several times in this book, and to his phrase *everywhen*. Two books of his are *Turtle Island* (New Directions, 1975) and *No Nature* (Pantheon, 1992).

Nancy Wood: An incredibly versatile poet, photographer, novelist, and naturalist, she penned two works I especially admire: *Many Winters* (Doubleday, 1974) and *Spirit Walker* (with paintings by Frank Howell) (Doubleday, 1993).

Sacred Architecture

Christopher Day and his book *Places of the Soul: Architecture and Environmental Design as a Healing Art*. London: Aquarian Press, 1993. In U.S.

Louis I. Kahn and his work in *Between Silence and Light: Spirit in the Architecture of Louis I. Kahn*. John Lobell. Boston: Shambhala, 1979.

Institute for Sacred Architecture, P. O. Box 556, Notre Dame, IN 46556-0556; (574) 232-1782; *www.sacredarchitecture.org*. Specializes in churches and produces the journal *Sacred Architecture*; *editor@sacredarchitecture.org*.

Institute for Studies in Sacred Architecture (ISSA), 9402 South 47th Place, Phoenix, AZ 85044; (480) 783-8787; *info@issarch.org*; *www.issarch.org*. Programs, projects, conferences, publications.

A. T. Mann and his book, *Sacred Architecture*. Rockport, MA: Element Books, 1993.

The Art of Right Placement

There are over seven hundred names used to describe the various forms of life energy. Some of those names are *prana, chi, the odic force, mana, noetic energy, tachyon, élan vital, bioplasma,* and *the Holy Spirit*. Regardless of how termed, the art of right placement concerns itself with how to build and furnish structures in accordance with the natural flow of such energy.

American School of Geomancy, (707) 544-8203; connect now at *rfa@richardfeatheranderson.com*; *www.richardfeatheranderson.com*. Based on Anderson's work, this school teaches the Americanized version of Chinese geomancy (feng shui). Comprehensive experiential programs are provided for in-depth training in the principles and practice of "the art of enhancing the sense of place and well-being and living in harmony with earth's patterns."

Creating Sacred Space with Feng Shui. Karen Kingston. New York: Broadway Books, 1997.

Feng Shui: Back to Balance. Sally Fretwell. Novato, CA: New World Library, 2002.

Feng Shui That Makes Sense. Cathleen McCandless. Minneapolis, MN: Two Harbors Press, 2011.

International Institute for Bau-Biologie and Ecology, P.O. Box 64188, Tucson, AZ 85728; 1-800-960-0333; *infopod@buildingbiology.*

net; *http://hbelc.org*. Bau-Biologie (a German term) is an "ecology-oriented value system." It came about out of concern for the health effects of people's surroundings. The system addresses selection of proper building sites, house design, energy aspects, heating/ventilation/air filtration, electrical installation, selection of proper building materials, light/illumination/color, furniture and interior design. Comprehensive.

Life Net Home, Gail England, 421 Bliss Pond, Calais, VT 05648; *amichieli@hotmail.com*; *info@lifenethome.org*; *www.lifenethome.org*. A group of people dedicated to right placement, sacred landscapes and building styles, and working with nature and natural forms for a healthy environment. Incorporates spiritual art into designs.

North American School of Geomancy, 610 Main Street, Great Barrington, MA 01230; (413) 528-8233; *geomancy@steinerbooks.org*; *www. steinerbooks.org*. Part of the Rudolf Steiner programs offered through the Steiner school system. Books on the subject available through their bookstores.

Sustainability

EcoVillage Training Center, P.O. Box 90, Summertown, TN 38483; (931) 964-4474; *www.thefarm.org/etc/*. Center for classes, workshops, and demonstrations of permaculture (a conscious design of cultivated ecosystems that have the diversity, stability, and resilience of nature). They teach that people's needs can be integrated into the landscape in such a way that will improve richness, beauty, and productivity.

William McDonough, leader in the field, can be contacted through University of Virginia Darden School of Sustainability, P. O. Box 6550, Charlottesville, VA 22906-6500; *herze@darden.virginia.edu*; *www.darden.edu/ web/about/business-perspective/sustainability/home/*. McDonough believes in zero emissions and buildings as a living, breathing part of nature. To reach him directly, access *www.mcdonough.com*. His book: *Cradle to Cradle: Remaking the Way We Make Things*, William McDonough and Michael Braungart. New York: North Point Press, 2002.

Orgone Biophysical Research Lab, Green Springs Center, P.O. Box 1148, Ashland, OR 97520; (541) 552-0118; *demeo@mind.net, www.orgonelab.org*. Following up on the breakthrough social and environmental research of the late Wilhelm Reich, James DeMeo has extensively researched cross-cultural relationships between climate and human behavior, in historical and geographical contexts—including practical methods to reverse the global trend towards desert expansion. His work is revolutionary. Will travel; query about publications and plans to build a research and educational center.

The Green Intention: Living in Sustainable Joy. Sandy Moore and Deanna Moore. Camarillo, CA: DeVorss & Co., 2011.

Permaculture: A Designer's Handbook. Bill Mollison. Australia: Tagari, 1988. In U.S. A classic.

Secrets of the Soil. Christopher Bird and Peter Tompkins. New York: HarperCollins, 1990.

Permaculture Institute, P.O. Box 3702, Santa Fe, NM 87506; (505) 455-0514; *www.permaculture.org*. Extensive offerings. Inquire.

Virginia Association for Biological Farming (VABF), P.O. Box 1003, Lexington, VA 24450; (540) 463-6363; *http://vabf.org*. One of the first statewide organizations in the nation to support sustainable agriculture. Membership based, they publish a newspaper and sponsor conferences and workshops. Along this same line, also contact *Community Farm Alliance*, 119½ West Main Street, Frankfort, KY 40601; (502) 223-3655; *www.communityfarmalliance.org* and *Southern Sustainable Farming*, P. O. Box 1552, Fayetteville, AR 72702; (479) 251-8310; *www.ssawg.org*. These two groups cover over thirteen states.

Water Research Institute of Blue Hill, 16 Ledge Road, #72, Blue Hill, ME 04614; (207) 374-2405; *jgreene@waterresearch.org*; *www.waterresearch.org*. These people research and construct flowforms. A flowform is a stylized, artificial watercourse that helps water to regenerate by introducing oxygen—so that aerobic bacteria can break down pollutants. A technology based on nature and imbued with art, it has many applications. For instance, flowforms in Sweden transport and aerate waste water in a natural purification system. In Amsterdam, flowforms humidify the air and create a refreshing acoustical background in a bank building. Wonderful for gardens. Flowforms cause vortexes that work similar to a partial colloidal condition in how standing or flowing water can be spun around to reinvigorate not only the water but also the environment.

Malcolm Wells Underground Art Gallery, 673 Satucket Road, Brewster, MA 02631; Fax 508-896-5116; *www.malcolmwells.com*. Now run by his widow Karen North Wells, Malcolm's Gallery is noted for detailed information about earth-sheltered architecture. A unique treasure trove of leading-edge ideas and designs.

Earth Energies

Earth Energies: A Quest for the Hidden Power of the Planet. Serge Kahili King. Wheaton, IL: Quest Books, 1992.

Earth Memory: Sacred Sites—Doorways into Earth's Mysteries. Paul Devereux. St. Paul, MN: Llewellyn Press, 1992.

Earth Mind: Communicating with the Living World of Gaia. Paul Devereux, John Steele, and David Kubrin. Rochester, VT: Destiny Books, 1989.

Mystery Teachings from the Living Earth: An Introduction to Spiritual Ecology. John Michael Greer. San Francisco: Weiser Books, 2012.

Shamanism and the Mystery Lines: Ley Lines, Spirit Paths, Shape Shifting and Out-of-Body Travel. Paul Devereux. St. Paul, MN: Llewellyn Press, 1993.

Sacred Geometry

The Dimensions of Paradise: Sacred Geometry, Ancient Science, and the Heavenly Order on Earth. John Michell. Rochester, VT: Inner Traditions, 2008.

Divine Proportion: The Mathematical Perfection of the Universe. Scott Olsen. New York/Somerset, United Kingdom: Alexian Limited, 2012. In U.S.

How the World Is Made: The Story of Creation According to Sacred Geometry. John Michell with Allan Brown. Rochester, VT: Inner Traditions, 2012.

Sacred Geometry: Deciphering the Code. Stephen Skinner. New York: Sterling, 2006.

The Vesica Institute, 1011 Tunnel Road, Suite 200, Asheville, NC 28805; (828) 298-7007; *info@vesica.org*; *www.vesica.org*. Holistic studies with a full range of classes and talks on Egyptian biogeometry, earth energies, sacred geometry, vibrational energy balancing, and more.

Different Patterns of History

To quote Kevin Todeschi from the Association for Research and Enlightenment (A.R.E.): "When enough individuals combine to utilize their free will in a positive direction, the potential future 'seen' by a psychic is altered. An excellent example of this is the Old Testament story of Jonah. He foresaw the destruction of Nineveh because of the people's own negative thoughts and deeds. God Himself was not going to destroy the city, rather it was the people themselves who were bringing about a destruction based upon their own selfishness and acts of hate toward one another. In order to 'help' the people, Jonah predicted the city's downfall. As a result, the people banded together, changed their thoughts and actions, and—through their unified free will—altered the foreseen future in less than forty days. The destruction never came, and Nineveh was saved."

We the people can change the future, and we do that by changing ourselves and the way we think.

Rethinking Our Future

James Burke, an example of an innovator. He produced *The Web*, a ten-part television series on the Learning Channel in 1994 (a sequel to his 1978 *Connections* series). His theory of learning webs is an attempt to explain the chain or relationships linking one part of life, in any century, to another. "Imagine knowledge and history as being rather like a three-dimensional globe made up of millions and millions of threads interacting across it. The center of the globe is the beginning, and the edge of the globe is the modern world." In his television series, he travels the web, showing how seemingly unrelated people, events, and ideas link up. These unexpected linkages illustrate broader patterns of history.

Example: the link between a reporter's tape recorder and malaria. "The ferrite recording tape was made by Germany's BASF, the Aniline and Soda Factory of Bavaria, a company that came into existence because an Englishman named Perkins, looking for a malaria cure in 1857, dumped an experiment in the sink and saw what it did to the water. He invented the world's first aniline dye, and in true British fashion, made his money by selling the idea to a German . . . the people he sold it to became BASF."

The Optimist Magazine, formerly ODE Magazine. This world-wide publication is dedicated to building a community that jumpstarts change. The articles they carry and the activities they sponsor inspire and teach how-to strategies that actually work. Contact: *www.ode.com*. U.S. office: 268 Bush Street #4419, San Francisco, CA 94104; *ode@odenow. com*. Netherlands Office: P.O. Box 2402, 3000 CK, Rotterdam, The Netherlands.

Starhawk, a teacher of collaboration. She authored *The Empowerment Manual: A Guide for Collaborative Groups* (New Society Publishers, 2011). Her website is *www.starhawk.org*. She is an activist with a long history of working with group dynamics, collective decision making, and dealing with difficult people.

Bill Strickland, a genius who sees the genius in everyone. He discovered the secret of how to inspire others to learn and to excel, and how to build schools in the most unlikely places. With the theme environment drives behavior, he has turned broken-down, dangerous areas into communities that care. He believes art is essential—and if you treat people better, they'll do better. His book with Vince Rause is *Make the Impossible Possible: One Man's Crusade to Inspire Others to Dream Bigger and Achieve the Extraordinary* (Broadway Books, 2007). Contact him at Manchester Bidwell Corporation, 1815 Metropolitan Street, Pittsburgh, PA 15233; (412) 323-4005; *wstricklandjr@mcg-btc.org*; *www.manchesterbidwell.org*.

Different Views on History (Both Practical and Profane)

America B.C.: Ancient Settlers in the New World. Barry Fell. New York: Pocket Books, 1976.

Catastrophobia: The Truth behind Earth Changes. Barbara Hand Clow. Rochester, VT: Bear & Co., 2001.

Forbidden Archaeology: The Hidden History of the Human Race. Michael A. Cremo and Richard L. Thompson. Badger, CA: Govardhan Hill, 1993.

Genesis Revisited: Is Modern Science Catching Up with Ancient Knowledge? Zecharia Sitchin. Santa Fe, NM: Bear & Co., 1990.

Human Devolution: A Vedic Alternative to Darwin's Theory. Michael A. Cremo. Mayapur, India: Torchlight Publishing, 2003. In U.S.

Jochman's Alma Tara Multi-Versity. P. O. Box 10703, Rock Hill, SC 29731-0703 (please write). An author and earth mysteries explorer, Dr. Joseph Robert Jochmans has distinguished himself in research concerning lost civilizations. He has a series of books on his work and offers educational materials and guided world tours. Inquire.

Lost City Series. David Hatcher. Five books: *South America,* 1986; *Africa and Arabia,* 1987; *Lemuria and the Pacific,* 1987; *China, Central Asia, and India,* 1987; *North and Central America,* 1992. Available from Adventures Unlimited, P. O. Box 22, Stelle, IL 60919.

Occult America: White House Séances, Ouija Circles, Masons, and the Secret History of How Mysticism Shaped Our Nation. Mitch Horowitz. New York: Bantam Books, 2010.

Presence: Human Purpose and the Field of the Future. Peter M. Senge, C. Otto Scharmer, Joseph Jaworski, and Betty Sue Flowers. New York: Crown Business, 2008.

Reimagination of the World: A Critique of the New Age, Science, and Popular Culture. David Spangler and William Irwin Thompson. Santa Fe, NM: Bear & Co., 1991.

Sacred Number and the Origins of Civilization: The Unfolding of History through the Mystery of Number. Richard Heath. Rochester, VT: Inner Traditions, 2007.

Spiritual Politics: Changing the World from the Inside Out. Corinne McLaughlin and Gordon Davidson. New York: Ballantine Books, 1994.

The Civilization of the Goddess. Marja Gimbutas. San Francisco: Harper-One, 1991. A classic.

An Intimate History of Humanity. Theodore Zeldin. New York: HarperCollins, 1994.

The Orion Mystery: Unlocking the Secrets of the Pyramids. Robert Bauval and Adrian Gilbert. New York: Crown, 1994.

The Secret Vaults of Time: Psychic Archaeology and the Quest for Man's Beginnings. Stephan A. Schwartz. New York: Grosset and Dunlap, 1978.

The Source Field Investigations. David Wilcock. New York: Dutton, 2011.

Worlds before Our Own. Brad Steiger. San Antonio, TX: Anomalist Books, 2007.

"Now is the time when we must renew ourselves and live as if we and all of life are sacred."
—Jean Houston

Notes

The longest journey is the journey inwards of him who has chosen his destiny—who has started upon his quest for the source of his being.
—Dag Hammarskjöld

1. Sacred geometry is a language of harmony, beauty, order, and balance. The labyrinth, as a part of that tradition, is a mind changer—spinning the one who treads its pathways from present brain development to enhancements of brain potential. Unlike a maze, the labyrinth is not designed to confuse, but to transform—*through the renewing of your mind.* As a tool of transformation, labyrinths gently simulate all the stages of colloidal states to bring one back into alignment with his or her highest good. Their job is to shift energy and uplift consciousness. That's why this book was turned into a labyrinth. It is not enough to talk about transcended states or about the spiritual journey. What is important is experience, that each person has a chance to get a sense of what a brain shift might feel like—sort of a teaser for the real thing. Thus, *Future Memory* is not a book. It is a device, a labyrinth. Enjoy your adventure! Here are some books and a company to help you understand labyrinths:

The Golden Age of Chartres: The Teachings of a Mystery School and the Eternal Feminine. René M. Querido. New York: Floris Books, 1987.
The Idea of the Labyrinth: From Classical Antiquity through the Middle Ages. Penelope Reed Doob. Ithaca, NY: Cornell University Press, 1990.
Labyrinths: Ancient Myths and Modern Uses. Sig Lonegren. Glastonbury, England: Gothic Image Publications, 1991.

Walking a Sacred Path: Rediscovering the Labyrinth as a Spiritual Tool. Lauren Artress. New York: Riverhead Books, 1995. Quoted in chapter 1 of this book.

The Labyrinth Company: 51 Summit Road, Riverside, CT 06878-2104; 1-888-715-2297; *info@labyrinthcompany.com*; *www.labyrinthcompany.com*.

2. The quotes about George Washington Carver were obtained from *The Woodrew Update*. In its time, this newsletter was one of the most practical, yet innovative, in the field of "New Thought," put together by Greta Woodrew (now deceased) and her husband Dick Smolowe, both L.L.D.s and successful in mainstream business. Woodrew authored *On a Slide of Light* (Macmillian, 1981) and *Memories of Tomorrow* (Dolphin, a division of Doubleday, 1988).

3. Findhorn Foundation, Cluny College, the Accommodations Secretary, Forres, Pineridge/The Park, Findhorn, Moray, Scotland, IV36 3TZ; *enquiries@findhorn.org*; *www.findhorn.org*. They have numerous publications including the book, *The Findhorn Garden: Pioneering a New Vision of Humanity and Nature in Cooperation*. Ask about catalogue, classes, and activity schedules.

4. Perelandra Center for Nature Research, P.O. Box 3603, Warrenton, VA 20188; *www.perelandra-ltd.com*. Question Hotline: (540) 937-3679, answered on Wednesdays (10:00 a.m. to 8:00 p.m.). Ask for catalogue and activity schedules.

5. *Indian Running*. Peter Nabokov. Santa Barbara, CA: Capra Press, 1981.

6. *The Reality Game and How to Win It*. Brad Steiger. Available from Newcastle Publishing Co., P.O. Box 7589, Van Nuys, CA 91409.

7. The amazing story of Friedrich Jergenson was obtained from the late George Meek's quarterly newsletter *Unlimited Horizons*, a publication that was devoted to the Electronic Voice Phenomenon. Of interest on this subject is the book *Voices of Eternity*, by Sarah Wilson Estep (Fawcett Books, 1988). Estep is founder and past-president of the American Association of Electronic Voice Phenomenon. Active today in the field are Tom and Lisa Butler, P.O. Box 13111, Reno, NV 89507; *aaevp@aol.com*; *http://aaevp.com*.

8. The incredible story of T.L. was obtained from an article entitled "Research Report—Psychic Leap-Frog by Automobile," written by Frank C.

Tribbe. This appeared in the Spring 1988 issue of *Spiritual Frontier Quarterly Journal* (Vol. 20, No. 2)—a journal no longer in publication.

9. The Door to the Secret City, by Kathleen J. Forti, tells in story form what it is like for a child to have a near-death experience and then deal with the aftereffects. Available now as a free offering online. Contact Kathy at *kjforti@aol.com* or through her company Trinfinity at *kathy@trinfinity8.com*.

10. Elisabeth Kübler-Ross, M.D., passed away in 2004. She leaves behind a legacy of work unparalleled in the field of death and dying. It was my privilege to have personally known her and to have been a guest in her home many times, when she lived in the western part of Virginia.

11. Wabun Wind has written many books, including her own story, *Woman of the Dawn: A Spiritual Odyssey* (Prentice Hall, 1989). The Bear Tribe Medicine Society is now Panther Lodge Medicine Society. After the passing of Sun Bear, Wind Daughter was chosen to continue the Native American teachings so much a part of Sun Bear's work, and that of other great Medicine Keepers. Contact: Panther Lodge Medicine Society, P.O. Box 2388, Mountain View, AR 72560-2388; (870) 368-7877; *winddaughter@centurytel.net*; *www.winddaughterwestwinds.org*; or *www.thebeartribe-medicinesociety.org*.

12. Coming Back to Life: The After Effects of the Near-Death Experience. P. M. H. Atwater, L.H.D. Dodd, Mead and Co., 1988. Now in print through Transpersonal Publishing, Kill Devil Hills, NC, 2008.

13. Diane Pike and Arleen Lorrance are co-founders of the Love Project organization (renamed Teleos Institute), and have moved their headquarters from San Diego, California, to Scottsdale, Arizona. The Love Project Principles originally came to Arleen in a burst of radiance when she was desperately seeking a way to help stem recurring violence in a ghetto high school. Not only were the Love Project Principles instrumental in transforming the high school, but through the steady guidance of Diane and Arleen, thousands of other people have also benefited from them. The session I attended with these two in Roanoke had as its text the book *Toning, the Creative Power of the Voice* by Laurel Elizabeth Keyes (DeVorss and Co., 1980). For further information about their current projects, contact: Teleos Institute, 7119 E. Shea Blvd., Suite 109, PMB 418, Scottsdale, AZ 85254-6107; (480) 948-1800.

14. James Van Avery's professional paper, as well as mine, was published in *Proceedings Book of the International Conference on Paranormal Research*, June 1989. Permission for me to use parts of Van Avery's paper was granted both by Van Avery and the conference staff.

15. *Flow: The Psychology of Optimal Experience*, Mihaly Csikszentmihalyi. New York: Harper & Row, 1990.

16. *Wildfire Magazine* (formerly *Many Smokes Magazine*), a publication of Earth Vision. This magazine is no longer printed. Refer to endnote 11.

17. *Cosmic Consciousness* by Richard Maurice Bucke, M.D., was first published in book form by Innes and Sons in 1901 although it was completed in manuscript form the year before. There are various reprint editions. The one I have is from Citadel Press, Secaucus, NJ, 1982.

18. *New Scientist Magazine* (January 8, 1994) covered the latest findings of Nicholas Humphrey, a senior research fellow at Cambridge, who discovered that emotions are primary. His work concerns "sensor consciousness," a name he coined for the brain's role in feeling. Other researchers have joined in, each adding more information about the importance of emotion and how it influences the mind. A good book on this subject is *Descartes' Error: Emotion, Reason, and the Human Brain* by Antonio R. Damasio (Grosset/Putnam, 1994).

19. For most of my childhood, I was guided along and taught by Spirit Keepers in the canyons and deserts of southern Idaho. I experienced Spirit Keepers as transparent beings of radiant light, whom I could clearly see and easily converse with. They told me their job in creation was and still is to maintain the integrity of spirit in matter so that form can exist (parallel to the function of lithium in that it bonds spirit with matter—refer to the paperback edition of *Beyond the Light* for a section on lithium, beginning with page 257). Other people regard these radiant beings as godlike and have given them fanciful names, but to me, they were more like beloved elders than anything reverent.

20. *Stalking the Wild Pendulum: On the Mechanics of Consciousness* and *A Cosmic Book: On the Mechanics of Creation*, both by Itzhak Bentov, are now under reprint rights with Destiny Books (Inner Traditions). A video by the author's widow, Mirtala, explaining his cosmology is available: *Itzhak Bentov, from Atom to Cosmos, Evolution of Consciousness as a New Model of the Universe*. To obtain a copy or seek further information about Bentov's work,

write to Mirtala, 10780 E. Oak Creek Valley Drive, Cornville, AZ 86325. The time-space diagram used here is from *A Cosmic Book* and appears with permission. When you contact Mirtala, also ask about books and videos on her transformational sculpture and her artwork. Mirtala was helpful with the original concept for this book, and for that I say, "Thank you."

21. "Conceptual Model of Paranormal Phenomena," by Jack Houck and dated September 21, 1982, was first published by *Archaeus Magazine*, Vol. 1, No. 1, Winter 1983. The same paper also appeared in the *American Dowser Quarterly Digest*, Vol. 34, No. 4, Fall, 1994; copies available through the American Society of Dowsers, 184 Brainherd Street, P.O. Box 24, Danville, VT 05828-0024; (802) 684-3417. The space-time unit diagram in appendix II came from Houck's "Conceptual Model." I thank Jack Houck for permission to use the diagram and for the time he spent explaining his model of paranormal phenomena to me.

22. *Alternative Realities: The Search for the Full Human Being*. Lawrence LeShan. New York: Ballantine Books, 1976.

23. *Vital Lies, Simple Truths: The Psychology of Self-Deception*. Daniel Goleman, Ph.D. New York: Simon and Schuster, 1985.

24. *The White Hole in Time*. Peter Russell. San Francisco: HarperOne, 1992. Also by Russell: *The Global Brain Awakens: Our Next Evolutionary Leap*. Palo Alto, CA: Global Brain, Inc., 1995.

25. *Hammond Barnhardt Dictionary of Science*. Robert K. Barnhart with Sol Steinmetz. Maplewood, NJ: Hammond, 1986.

26. *Black Holes*. John G. Taylor. New York: Avon Books, 1973.

27. *A Brief History of Time*. Stephen Hawking. New York: Bantam Books, 1988.

28. *Beyond the Light: What Isn't Being Said about the Near-Death Experience*. P. M. H. Atwater, L.H.D. New York: Birch Lane Press, 1994. Now in print through Transpersonal Publishing, Kill Devil Hills, NC 2009.

29. The legacy of Edgar Cayce is preserved by the Association for Research and Enlightenment (A.R.E.), 67th and Atlantic Avenue, Virginia Beach, VA 23451; 1-800-333-4499; *are@edgarcayce.org*; *www.EdgarCayce. org*. This nonprofit organization houses a large library containing a host of materials on parapsychology, metaphysics, mysticism, and health. They also offer a full range of services, publications, and learning opportunities

devoted to spiritual development and wholistic lifestyles. Study groups like the one I attended are now quite numerous, even internationally. Ask about study groups near you when you call, and request a copy of their publications catalogue.

30. *Stephen Hawking's Universe: An Introduction to the Most Remarkable Scientist of Our Time.* John Boslough. New York: Avon Books, 1989.

31. I was first introduced to this concept through various shamanistic traditions and from my own study of ancient cultures. Then I read the work of Issac N. Vail, printed in *Stonehenge Viewpoint*, a bimonthly publication, particularly his article "The Crystal Veil," which appeared in the January/February 1987 issue. His research on the theory of a vapor canopy said to have once covered prehistoric skies around the earth is impressive. There are now a number of sources available on this intriguing topic. Most of them can be traced through bibliography references, using the topic Issac N. Vail and the Vapor Canopy Theory.

32. Books by Harold Saxton Burr, Ph.D., were published in England. His most noted works are *Fields of Life* and *Blueprint for Immortality*. It is impossible in this book to give proper credit to Burr for his pioneering work or even to adequately recognize his genius. It is possible, however, that by mentioning him here, others will be inspired to further investigate his work. Refer to the article "The Electrical Patterns of Life: The Work of Dr. Harold S. Burr" by WRF. Available at *www.wrf.org/men-women-medicine/dr-harold-s-burr.php*.

33. Books by Rupert Sheldrake: *A New Science of Life: The Hypothesis of Formative Causation.* Los Angeles: J. P. Tarcher, 1981; and *The Presence of the Past: Morphic Resonance and the Habits of Nature.* New York: Times Books, 1989.

34. *Modern Aether Science.* Harold Aspden. Order directly through Sabberton Publications, P.O. Box 35, Southhampton, England (ISBN 0850560039).

35. *Children of Light* depicts the holy sparks of God coming to earth to experience the power of choice and then returning to their Source, wiser than before. The children's storybook was written and illustrated by Jan Royce Conant. Obtain directly from the author at Stonefield Farm Studio, Three Bridges Road, East Haddam, CT 06423; (203) 434-9030.

36. *The Silmarillion*. J. R. R. Tolkien, edited by Christopher Tolkien (Boston: Houghton Mifflin, 1977), in conjunction with George Allen and Unwin Ltd., the original publishers in England. Tolkien also wrote *The Lord of the Rings* and *The Hobbit*, as well as numerous other books and papers on language origins and myth.

37. James Lovelock produced the following books on the Gaia Hypothesis: *Gaia: A New Look at Life on Earth*. University Press, 1979; and *The Ages of Gaia: A Biography of Our Living Earth*. W. W. Norton, 1989.

38. *The Power of Limits: Proportional Harmonies in Nature, Art and Architecture*, György Doczi. Boston: Shambhala, 1985. Quotations used in this book are with permission from the publisher.

39. *Head First: The Biology of Hope*. Norman Cousins. New York: E. P. Dutton, 1989.

40. *The Body Electric: Electromagnetism and the Foundation of Life*. Robert O. Becker, M.D., and Gary Selden. New York: William Morrow, 1985. Quotes used in this book are with permission from the publisher.

41. *Between Silence and Light, Spirit in the Architecture of Louis I. Kahn*. John Lobell. Boston: Shambhala, 1979.

42. *Newsweek* Magazine subscription department, *www.newsweeksubscriptions.com*. Back issues are available on microfilm at most libraries across the country; ask for archival department, or check at *http://magazine-directory.com*.

43. Vernon M. Sylvest, M.D., no longer operates the Richmond Health and Wellness Center. He authored *The Formula*, a book about someone who gets sick and gets well and why. It is available from Sunstar Pub., Ltd., Fairfield, IA, or through the author. The story "The Wasp Did Not Sting" was adapted from the July 1989 newsletter, formerly published by the Institute for Higher Healing. Dr. Sylvest was kind enough to allow an edited version of that story to appear in this book. He and his wife, Anne, bought the Elisabeth Kübler-Ross property near Head Waters, Virginia. They have converted it into a holistic health retreat called Healing Waters Lodge. Contact them at: Healing Waters Lodge, 714 Healing Waters Drive, Head Waters, VA 24442; (540) 396-3466; *office@healingwaterslodgevirginia.com*; *www.healingwaterslodgevirginia.com*.

44. For further information about Cleve Backster and the incredible research he has done, contact the Backster School of Lie Detection, (619)

233-6669; *clevebackster@cs.com*; *www.backster.net/*. If you Google his name, you will find a rich source of videos and films about his work and the difference his research has made in our understanding of "primary perception."

45. *Prominent American Ghosts*. Susy Smith. New York: World Publishing, 1967.

46. *Abraham Lincoln, The Prairie Years and War Years*. Carl Sandburg. New York: Harcourt Brace Jovanovich, 1954.

47. *Mysteries of the Unknown*. Time-Life Books. New York: Time-Life-Books, 1987.

48. *The Cosmic Code*. Heinz R. Pagels, Ph.D. New York: Bantam Books, 1984.

49. *Conversations with Nostradamus*. Dolores Cannon. Huntsville, AK: Ozark Mountain, 1992. A three-volume set.

50. *A Brief History of Time*. Stephen Hawking. New York: Bantam-Books, 1988.

51. *The Looking Glass Universe*. John P. Briggs, Ph.D., and F. David Peat, Ph.D. New York: Cornerstone [Simon and Schuster], 1984.

52. References to Lincoln:

The Early Life of Abraham Lincoln. Ida M. Tarbell. New Brunswick, NJ: A. S. Barnes and Co., 1974.

The Hidden Lincoln, from the Letters and Papers of Wm. H. Herndon. Emanuel Hertz. New York: Blue Ribbon Books, 1940; Viking Press, 1938.

The Intimate Lincoln. Joseph E. Suppiger. Lanham, MD: University Press of America, 1985.

The Life of A. Lincoln: From His Birth to His Inauguration as President. Ward H. Lamon. Boston: James R. Osgood and Co., 1872.

The Lincoln Nobody Knows. Richard N. Current. New York: McGraw Hill Books, 1958.

Lincoln: A Psycho-Biography. L. Pierce Clark. New York: Charles Scribner's Sons, 1933.

53. *Closer to the Light, Learning from the Near-Death Experiences of Children*. Melvin Morse, M.D., with Paul Perry. New York: Villard Books, 1990; and *Transformed by the Light*. Melvin Morse, M.D., with Paul Perry. New York: Villard Books, 1992.

54. *Einstein: The Life & Times*. Ronald W. Clark. New York: Thomas Y. Crowell Co. [World Publishing Co.], 1971; and *Subtle is the Lord. . . : The Science & The Life of A. Einstein*. Abraham Pais. Oxford, U.K.: Oxford University Press, 1982.

55. *Mozart*. Hugh Ottaway. London: Salem House Orbis Publishing, 1979.

56. *Chaos: Making a New Science*. James Gleick. New York: Viking, 1987; *Nature's Chaos*. James Gleick and Eliot Porter. New York: Viking, 1990; *A Turbulent Mirror: An Illustrated Guide to the Chaos Theory and the Science of Wholeness*. John Briggs and F. David Peat. New York: Harper & Row, 1989.

57. *Powershift*. Alvin and Heidi Toffler. New York: Bantam Books, 1990.

58. *The Game of Life and How to Play It*. Florence Scovel Shinn. New York: Fireside, 1986. Other books from Shinn: *Your Word Is Your Wand* and *The Power of the Spoken Word*, both through DeVorss and Co., Marina del Rey, CA 1978.

59. The May 7, 1993, edition of the *Seattle Times* newspaper carried an article about a new compound called Calanolide A. *Calophyllum lanigerum*, a member of the Guittiferae family of gum-producing trees, was found in a swamp in the Malaysian province of Sarawak. Biologists had collected two pounds of twigs, bark, and fruit from the trees to experiment with, then tested the substance they found (Calanolide A) against HIV-1 virus. The result? It was 100 percent effective in blocking the virus. Test details were published in the 1992 *Journal of Medical Chemistry*. When the biologists returned to Malaysia several years later for more samples, they discovered that the specific trees they wanted had been cleared from the swamp by local residents needing more land. Identical trees in surrounding areas did not yield Calanolide A, nor have they had any luck finding another source at this writing.

60. My *The Magical Language of Runes* (Bear & Co., 1990); *Goddess Runes* (Avon Books, 1996); and *Runes of the Goddess* (Galde Press, 2007) cover the yin or elder runes, the feminine system of rune use. "The Way of a Cast," using Goddess Runes, is what helped me to retrain my brain after my near-death experiences in 1977. I wrote *The Magical Language of Runes* as a way of saying thank you to the runes, and then passing the skill on to others. (*Goddess Runes* and *Runes of the Goddess* are expansions of that original text).

61. Dean Black's concepts about health and wellness were the most sensible and compelling I had yet heard. His talks and workshops were geared to help people realize that one must address the whole, not just part of the whole, to establish a true context for healing. His ideas challenged the entire medical establishment to reconsider their present model of what constitutes health. He wrote *Health at the Crossroads: Exploring the Conflict between Natural Healing and Conventional Medicine* (Tapestry Press, 1988).

Along the line of contextual healing is this interesting note: a double-blind study assumes that the mind has no effect on outcomes, when the opposite is true. This means that a double-blind study can actually distort results to the point that the reverse of what is actually happening appears to be true. Double-blind studies, therefore, are unreliable in many instances of research, especially where the human body and medicine are concerned. The body is the instrument of the mind, not vice versa. We see this clearly with cancer. Cancer is not a molecular disease, but a contextual one, dependent upon and primarily caused by a multiple arrangement of conditions, the breakdown of context (wholeness).

62. John White had had over fifteen books published in the fields of consciousness research and paranormal phenomena. His articles have appeared in such publications as the *New York Times, Saturday Review, Science Digest, Esquire, Omni, Woman's Day,* and *Reader's Digest.* Among his many books, *A Practical Guide to Death and Dying* (Theosophical Publishing House, 1980) is exceptional. This new version of the Pledge of Allegiance was quoted with White's kind permission.

Let the beauty we love be what we do.
—Jelaluddin Rumi

Index

illusions.55-56. *See also* time and space
meaning, 221-22
paradoxical, 95-104
perceptual, 200, 203-04
imagination, 81, 191, 196-97, 198-99
as real, 51-52
implosion. 247. *See also* black holes
in-between, 7, 12-13, 115, 122-25, 128-29, 146-47, 157, 258-59
individuations, 64
infinity, 64-65, 162-63
information. 83, 157, 203-04. *See also* knowing
global, 209
ways to experience, 214-215
Ingersoll, Charles W., 13-14
Intelligence, 64-65, 156, 157, 294-95
as extension of memory, 213
intention, 83, 149
international Association of Near-Death Studies (IANDS), 274-75
intuition, 213, 290-94

Jergenson, Friedrich, 13-14
Jesus, 142-43, 175, 287
Jung, Carl Gustav, 70, 75, 149, 216, 281-82

Kahn, Louis, 177-78
Kairos, 193
Karma, 225-27
Kellogg, Joan, 283-85
knowing, 148-51, 213. *See also* remembering
before vs. after believing, 222-23
Kubler-Ross, Elizabeth, 34-36, 39

Labyrinth, 5-6, 64-65, 130
Laird, George, 10-11
learning , 11-12, 114, 280-81, 284-85
as remembering, 83
LeShan, Lawrence, 7, 95, 113, 292
Life, 83, 230-31
as mirror, 224-26
life patterns, 285-86
light, 114-116, 150-152, 157, 162-63, 169-70, 176
limbic system, 79-82, 124-25, 208, 247
as gateway, 93, 113, 128, 248-52
limits and limitlessness, 174, 175, 178. *See also* infinity
Lincoln, Abraham, 196-98, 205-07, 316-17
Lorrance, Arleen, 43-44
Love, 153-54, 168, 226-27, 227, 232
Lovelock, James, 173-74, 210

magic, 214
Maslow, Abraham, 69
Matter, 114-16, 119, 164
as manifestation of information, 164
as solidified light, 114-16, 146-47
subjective side, 228
matter/energy/light, 114-118, 122, 140
McKnight, David, 42, 158-59
McKnight, John L., 306
Medicine, 175-76, 210-11, 277-78
Meditation, 46, 69, 71, 72-73, 159-60
memory banks, 149, 150-51
memory fields, 156, 157, 213
memory (ies), 63, 78, 82, 91-92, 158-59, 198-99
conceptions and types of, 83-84, 279
remembering, 84, 86-87, 126-27, 212
mental models, changing patterns in, 64
mercy, 217-18
metamorphosis, 251
Michelangelo, 177
Mind, 91, 192, 194, 248-49
higher, 210, 211
mirroring reality, 52
miracles, 16, 220-22
Morse, Melvin, 206-07
Mozart, 207
Mumford, Lewis, 245-47
Music, 83, 96, 170-71, 177, 207, 293-94
Mysticism, 249-50

Nabakov, Peter, 10-11
near-death experiences, 72-73, 108, 125-27, 193. *See also* Atwater, Phyllis M.H.; *Beyond the Light*; *Coming Back to Life*
of children, 63, 64, 205-07
discoveries in, 164
of famous persons, 205-07, 265-71
as rites of passage, 128
transformations in, 165
types, 126
near-death survivors, 42, 150-51, 163. *See also* Atwater
ability to "remember" future, 5
Newton, Isaac, 214
nitrous oxide (laughing gas), 19-21
normalcy, 89, 215
nuclear energy, 153, 154
numbers, 256-59

Ohno, Susumu, 177
Omnipresence, 90, 129
One/Oneness, 126, 164-66, 169, 170-71
One True Source, 167, 168, 169-70

Orgasm, 169
otherworld journeys, 62, 63, 141-42, 193
out-of-body experiences, 84-87

Pagels, Heinz R., 197-98
parenting , 280-82
Parker, Richard, 12-13
past life memories, 83
patience , 130-31, 132-33, 143-47, 226-27
patterns and shapes, 53-55, 75, 82
Pearce, Joseph Chilton, 280
Peat, F. David, 200-03
Penrose, Roger, 139-40
Perception, 10, 11, 12, 82, 94, 200-03,
    247-51. *See also* intuition
        as determining truth/reality, 7, 97
perceptual illusions, 101-02, 200. *See* also
    illusions
perceptual preference/prejudice, 95-100,
    128, 226-27
Perelandra Gardens, 9, 229-30
photocopy process, "received" from spirits,
    7-8
 physical sensations, 27, 67
Pike, Diane, 43-44, 311
Plants, 180-83, 210, 229-30
Pledge of Allegiance, new version, 229-30
Poe, Edgar Allen, 12
Poetry, 300
Potential, 251
power, sources of, 203-04, 222-23
precognition, 17, 19, 20, 24, 217
preliving future, 4, 22-24, 192. *See also*
    future memory
        purpose, 30-31
premonitions, 222-23
        historical examples, 196-99
present moment, as past tense, 24
Price, Alfred, 276-77
Prophesying, 17
psychic ability, 184-85, 213-14, 290-94
psychokinesis, 90-91, 242-43

quantum physics, 102-04

radiation, 153, 159
rape, 22-23, 217-220
Reagan, Nancy, 12-13
Reagan, Ronald, 12-13
"real," 3, 7, 97, 98
        vs. "unreal," 5, 11-12, 94-95
reality shifts, 11-15
reality(ies), 29, 52, 97, 113, 215
        deeper web of, 148-52
        experienced differently, 16
        facets of our, 173

multiple/alternate, 64-65, 95,
    199-200
receptivity, 90-91, 214, 222-23
reconnections, 167-68
rediscovery, 62
reincarnation, 83, 168-69
relativity, theory of, 115-16, 118, 154-55,
    162-63. *See also* black holes
relaxation, 29, 68, 83
religion, 208, 232, 249-50, 287-88
remembering , 5, 82, 126-27, 212
        of rememberings, 84, 86-87
Repp, Don, 45-46
Repp, Neddy, 45-46
Reproduction, 149, 169-70, 182-83. *See*
    also creation
Resonance, 75-77, 86-87, 150, 151, 152
responsibility, taking, 222-24, 225-26
reticular activating system, 97-100
revelations, 17. *See also* futuristic
    awareness
Robertson, Morgan, 12-13
Russell, Peter, 115, 227

Sadat, Anwar, 185-86
Self, 68, 86-87, 224-25, 274
self-deception, 281-84
self-development, 281-83
self-perception, reassessing, 106-08
selves, reunion of, 86-87
sequential ordering, 16, 29, 64, 79, 92-93
sex, 169
shadow, 216, 224-27
Shakespeare, 207-08, 263-71
Sheldrake, Rupert, 150-51
soul creation, 166
soul formation, 166
soul mates, 168
soul nucleus, 166-70
soul projections, 166-67
soul(s), 86-87, 167, 169-70, 200-01, 287,
    290
        "twin," 167-68
space. *See* time and space
Sperry, Roger, 202-03
spirit, 8-9, 156
        Pure Spirit, 156
spirit creation, 165-66
Spirit Keepers, 86-87, 312
Spirituality, 249-50, 287-290
Stereotypes, 63, 64
Storytelling, 283-85
Strangers in a Strange Land (Heinlein),
    11-13
Subconscious, 79, 81, 87-88, 90, 97, 149,
    247

About the Author

P.M.H. Atwater is one of the foremost researchers of the near-death phenomenon. Her first two books, *Coming Back to Life: The After-Effects of the Near-Death Experience* and *Beyond the Light: What Isn't Being Said about the Near-Death Experience*, are considered on par with the work of Raymond Moody and Kenneth Ring. Atwater says her latest book, *Children of the New Millennium*, "is a major challenge to the entire field."

The author was inspired to write a trilogy, of which *Future Memory* is the second book, during her third near-death experience.

"I was given my mission once I returned," she says. "I was told: Test revelation. You are to do the research. One book for each death.'"

P.M.H. Atwater holds many writing awards, among them recognition for outstanding writing from Idaho Press Women; has been named to a variety of American and international editions of *Who's Who* and other biographical listings; and has been the keynote speaker for the International Forum on New Science and International Association for Near-Death Studies conference at San Antonio.

She lives in Charlottesville, Virginia, where she pursues the hobby of cooking, which won her many prizes at Idaho county fairs. She also enjoys "spending time with my three children, four grandchildren, my husband, and God."

Hampton Roads Publishing Company

. . . for the evolving human spirit

Hampton Roads Publishing Company publishes books on a variety of subjects, including spirituality, health, and other related topics.

For a copy of our latest trade catalog, call (978) 465-0504 or visit our distributor's website at *www.redwheelweiser.com*. You can also sign up for our newsletter and special offers by going to *www.redwheelweiser. com/newsletter/*.